HOW DO YOU

Find an Exoplanet?

HOW DO YOU

Find an Exoplanet?

JOHN ASHER JOHNSON

PRINCETON UNIVERSITY PRESS

PRINCETON AND OXFORD

Published by Princeton University Press, 41 William Street,
    Princeton, New Jersey 08540
In the United Kingdom: Princeton University Press, 6 Oxford Street,
    Woodstock, Oxfordshire OX20 1TW

press.princeton.edu

ISBN 978-0-691-15681-1

Library of Congress Control Number: 2015953471

British Library Cataloging-in-Publication Data is available

This book has been composed in Garamond and Helvetica Neue

Printed on acid-free paper ∞

Typeset by S R Nova Pvt Ltd, Bangalore, India

Printed in the United States of America

10 9 8 7 6 5 4 3 2 1

In space there are countless constellations, suns and planets; we see only the suns because they give light; the planets remain invisible, for they are small and dark. There are also numberless earths circling around their suns, no worse and no less than this globe of ours.

—Giordano Bruno, *De l'infinito universo et Mondi* (1584)

Eventually, we reach the dim boundary—the utmost limits of our telescopes. There, we measure shadows, and we search among ghostly errors of measurement for landmarks that are scarcely more substantial. The search will continue. Not until the empirical resources are exhausted, need we pass on to the dreamy realms of speculation.

—Edwin Hubble, *The Realm of the Nebulae* (1936)

CONTENTS

PREFACE

I have to confess, when I was first asked to write a book about exoplanets I was doubtful whether there existed an audience that hadn't already been served. After all, in the past decade many excellent books have been written on the subject, with levels of sophistication ranging from popular science up through graduate-level textbooks. However, after a bit more thought I realized that I interacted with my potential audience on a daily basis. Many universities and research institutions offer vibrant summer research programs for undergraduates. Every year during my time at Caltech, and more recently at Harvard, I receive numerous requests to serve as mentor for students ranging from fresh-men to rising seniors. These students are drawn to the field of exoplanetary science for the same reason that exoplanets are so popular among the general public. Planets are strongly associated with childhood memories of favorite science fiction films and comics, and the existence and nature of planets around other stars are key to answering some of humankind's most ancient questions about our place in the Universe.

The students seeking to pursue exoplanet research typi-cally have strong physics backgrounds, but they often lack experience in astronomy, particularly in the subjects most germane to astrophysical research. What these students need is a resource that builds upon their existing physical intuition to form an understanding of how planets are discovered and studied. Thus, the question for young, would-be researchers is not *why* they should start studying planets, but *where* and *how* they should start their learning. When I am approached by prospective summer researchers, I usually give reading assignments from the scientific literature and from various graduate-level textbooks. However, tackling some of these background sources can prove to be daunting for many students, even those with a Caltech or Harvard physics education. It would be advantageous to have a single, accessible textbook that covers the fundamentals and provides students with a launching point into the astronomical literature and various higher-level texts.

Fortunately, the fundamental physics behind exoplanet detection is the same physics that students learn in their freshman and sophomore classes. One need not understand tensors and field theory to understand the effects of planets on their stars. This is not to say that exoplanet detection is trivial, and that the details of, say, orbital dynamics are necessarily simple. General relativity does play an important role in many studies of exoplanets, and the mathematics involved in finding and characterizing planets can reach very high levels of complexity. But these high-order effects need not be considered until a student is deep into her research project. At the outset, a first-order understanding of orbits and simplifying approximations to the geometry of planets around their stars is sufficient to build a strong physical intuition of every method that has thus far yielded exoplanet detections.

To maintain a high level of accessibility, I eschew detailed derivations in favor of simplified pictures that nonetheless retain the basic physics of the situation. For example, a detailed understanding the Doppler shifts of stars induced by the gravitational tug of planets requires starting from a statement of the vector positions of the masses involved and builds from there into the development of reduced masses and average positions along elliptical orbits. However, the essence of the physical picture is retained by considering two masses in circular orbits about their mutual center of mass. By removing eccentricity from the picture, pages of mathematics can be reduced to a few equations. The reduced mathematical complexity makes the basic physics of orbital motion, the mutual gravitational tug between a planet and star, more readily apparent. This is the approach that I take in teaching introductory astronomy for majors. I make the assumption that there will be plenty of time to fill pages of algebra later. But at the outset the focus should be on the basic physics rather than the detailed mathematics.

Another advantage of simplifying the physical picture is that it allows the student to easily approximate the size of the signals involved in planet hunting. By doing so, it becomes obvious why the first planets orbiting other stars were not discovered until a mere two decades ago. The signals that planets induce are absolutely tiny, and discerning these signals from the noise is a considerable technical challenge. It is one thing for an expert to say that planet detection is difficult. With this book I aim to enable students to start with Newtonian mechanics and arrive at that conclusion for themselves.

A more concrete goal of this book is to teach the reader to intuitively understand the physics behind a planet signal. To put it another way, I want the reader to know how to read the properties of a planet and its orbit from a figure in a discovery paper. In each chapter I will walk the reader through the process of converting features of the discovery signal—for example, the depth of a transit dip, or the amplitude and duration of a microlensing signal—into the physical properties of the planet and its orbit. Measuring a planet's mass to several significant digits based on a time series of a star's radial velocity requires a detailed model. However, a by-eye evaluation of a radial velocity plot presented in a journal article can give the mass to at least an order of magnitude, and to one significant digit with a little more effort, but without the need to fit a model.

By making the topic accessible to undergraduate physics students, I hope this book will also be useful for anyone with a technical background and an interest in exoplanets. Astronomy has a rich history of amateurs making significant contributions, and exoplanetary science is no exception. Many people have had their interest piqued by news articles on the latest planet detections and popular science books about exoplanets. It is my hope that this book will bridge the gap between popular science and the details behind the press releases.

As a practical matter, and in an attempt to maximize the utility of this book, I focus primarily on the four detection techniques that have thus far produced trustworthy, useful planet discoveries. These techniques are Doppler, transit, microlensing, and direct imaging (Chapters 2–5). This book is not meant to be exhaustive. For example, I do not provide a detailed description of the astrometry method, which despite its long history as a detection technique has not yet reached the level of success of the other methods. However, the physical concepts behind the Doppler technique are shared by the astrometry technique, and a student who understands the Doppler detection method can understand a star's motion across the sky caused by an orbiting planet. The same can be said about various timing techniques, such as the method used to find the first exoplanets orbiting a pulsar (Wolszczan & Frail, 1992).

The focus of this book is on answering the question posed in the title: *How do you find an exoplanet?* Thus, I do not cover the techniques used to characterize planetary systems using follow-up measurements. For example, once a planet is found to transit its star, one can measure the planet's atmospheric composition using additional transit measurements at different wavelengths, or the alignment between the planetary orbit and stellar spin axis through follow-up Doppler measurements made during transit. These types of measurements are near and dear to me as they make up a large portion of my group's research activities. However, these measurements are made *after* a planet has been found, and therefore lie beyond the scope of this book. For more on characterization techniques see Sara Seager's thorough, graduate-level textbooks *Exoplanets* and *Exoplanet Atmospheres*. I also recommend Michael Perryman's *Exoplanet Handbook*.

To enhance portions of the book I describe some of the history behind each detection technique. But my historical treatment is far from complete. For more on the history of exoplanets I recommend Michael Lemonick's *Other Worlds: The Search for Life in the Universe*, or in the astronomical literature I strongly recommend reading the introduction of Fischer et al. (2014), and review articles by Gaudi (2012), Walker (2012) and Oppenheimer & Hinkley (2009). For a less technical view of the present state of exoplanetary science I recommend *Strange New Worlds* by Ray Jayawardhana. For a more future-tense view of where the field is heading be sure to read Lemonick's *Mirror Earth: The Search for Our Planet's Twin*.

In the process of writing this book, I have learned a great deal. I have drawn upon the knowledge of many of the other planet hunters in the world, many of whom I am proud to call my collaborators and friends. These experts include my former thesis adviser and mentor Geoff Marcy, as well as Debra Fischer and Jason Wright. Since my primary competency is in the Doppler and transit techniques of planet detection, I relied on others for expertise in other methods. For the development of the chapter on direct imaging, I am indebted to my former postdoctoral researchers Justin Crepp and Sasha Hinkley. For the chapter on microlensing, I must give most of the credit to Scott Gaudi, Jennifer Yee and Rosanne Di Stefano. For transits, I learned most of what I know from Josh Winn. Special thanks to Owen Gingerich, Jonathan Swift, Yutong Shan, Andrew Vanderburg and Erin Johnson for their helpful suggestions and edits. Finally, I wish to extend my thanks to the anonymous reviewers whose feedback greatly improved this book.

I must also thank the encouragement, support and intellectual environment formed by my amazing team, the Exolab. The undergrads, grad students and postdocs in my group are my secret weapon in tackling the toughest questions in stellar astrophysics and planetary science.

HOW DO YOU

Find an Exoplanet?

1

INTRODUCTION

> For as long as there been humans we have searched for our place in the cosmos.
>
> — *Carl Sagan, 1980*

1.1 My Brief History

I am an astronomer, and as such my professional interest is focused on the study of light emitted by objects in the sky. However, unlike many astronomers, my interest in the night sky didn't begin until later in my life, well into my college education. I don't have childhood memories of stargazing, I never thought to ask for a telescope for Christmas, I didn't have a moon-phase calendar on my wall, and I never owned a single book about astronomy until I was twenty-one years old. As a child, my closest approach to the subject of astronomy was a poster of the Space Shuttle that hung next to my bed, but my interest was piqued more by the intricate mechanical details of the spacecraft rather than where it traveled.

Looking back, I suppose the primary reason for my ignorance of astronomy was because I grew up in a metropolitan area, in the North County of St. Louis, Missouri. The skies are often cloudy in the winter when the nights are longest, the evenings are bright with light pollution from the city even when the clouds are absent, and the air is humid and mosquito-filled in the summer.[1] Another important factor is that from age six until twelve I spent a good fraction of my time in my room building with Legos. From early on I seemed destined to be an engineer rather than the astrophysics professor I am today.

It wasn't until I attended a small engineering college in the town of Rolla, Missouri (pronounced locally as Rah-lah, Mizz-ur-ah) that I had my first memorable experience with the night sky (the school is now known as the Missouri University of Science & Technology). The town of Rolla boasts a population of about 20,000, but only during the academic year; once the students leave for summer break, the population dips below 12,000. One hot summer evening in August 1997, just before the start of the fall semester, I was sitting in my room playing computer games over our homemade local area network when my roommate Jason convinced me to go out that night to see the Perseids meteor shower. Jason had learned about the shower from a public service announcement from the student radio station, KMNR, where we both worked as deejays. With nothing much better to do before classes started, we and several friends drove out to a farmer's field just outside of town, threw out a few blankets and waited for the shooting stars.

As we watched the meteors of various sizes streak across the night sky, I noticed a swath of faint, splotchy light and I asked if anyone else saw it, too. It turned out that we were

---

[1] But how 'bout them Cardinals?!

seeing the Milky Way—our own galaxy as viewed from the inside—along with many of the summer constellations. It was that event, at the age of twenty-one, that sparked my interest in the grander Cosmos.

From that night on I started looking up to notice twinkling stars of various subtle hues, the phases of the Moon, and the sky in general. One night later that year I saw a remarkably bright star that I hadn't noticed previously. A few weeks later I asked one of my physics professors, Dr. Schmitt, about the mysterious star expecting to hear about some component of a constellation with a Latin name or some boring numerical designation. I explained about how much brighter it was than the surrounding stars, and how it seemed to stand out so much clearer than everything else in the sky. He smiled and told me that I wasn't seeing a star at all. Rather, I was seeing the planet Jupiter. Thinking back to that moment now, I suppose that was when I had just "discovered" my first planet! The discovery may have been thousands if not millions of years old and therefore hardly new to humanity. But it was new to me and it further sparked my interest in the subject of astronomy.

1.2 The Human Activity of Watching the Sky

Throughout history humans have taken inventory of the night sky, its stars, planets and other luminous bodies. Given many more nights staring up, rather than down at my textbooks or across at my computer screen, I would have noticed other bright planets in addition to Jupiter, including Mercury, Venus, Mars and Saturn. If I had paid careful enough attention, as my ancestors had thousands of years ago, I would have noticed that the planets did not always appear in the same place month after month with respect to the surrounding stars. Compared to the relatively static background of the constellations, the planets wander at their own pace, and sometimes move in a direction opposite of the stars. The activity of measuring the positions of astronomical objects relative to the background of relatively static stars is known as *astrometry*. The history of exoplanets starts with humans conducting astrometric measurements of the planets, and the story of the discovery of the nature of the Solar System planets marks the dawn of science as we know it.

While the word *planet*, meaning "wandering star," originates with the Greeks, astrometric measurements of planets were first recorded by the Babylonians, who started recording the positions of Venus sometime in the seventeenth century B.C.E. We know this based on the ancient Venus Tablets, on which these earlier astrometric observations were later reproduced and preserved. These tablets date back to the seventh century B.C.E., and the original observations were most likely made in relation to religious customs and beliefs, with the times of maximum separation from the Sun associated with various omens. Later Babylonian texts noted the positions of other planets, along with the Sun and the Moon, and noted the periodic nature of these objects' motion in the sky.

The Babylonian tabulations of planetary positions are the earliest written records of humans tracking the positions of night-sky objects. However, while various aspects of planetary motion were noted to be periodic and therefore predictable, ancient astronomers had no proper system of motion attributed to the planets and stars. With the advent of geometry, the Greeks later devised the first mathematical description of the movements of objects in the sky. They posited that the Earth was a sphere— they were aware that the Earth is not flat—and around the central sphere rotated a larger sphere containing the stars and planets. This "two-sphere" model was advanced by the most prominent philosophers of the time, notably Socrates, Aristotle and Ptolemy, and later adopted by Christian theologians through the sixteenth century B.C.E. (Kuhn, 1957).

The preference for the two-sphere model was not necessarily based on its ability to make accurate physical explanations of observed phenomena, but were instead based on assertions about nature that had aesthetic, rather than scientific appeal as we would demand today. According to the ideas of the early Greek philosophers, later expanded upon by Ptolemy, the basic elements of the Universe had preferred locations and behaviors. For example, in a Ptolemaic Universe, the Earth and its constituents—earth, water, air and fire—were changing and imperfect. Further, things that are of the Earth tend toward its center, which provides an explanation for why objects are pulled to the Earth. While things of the Earth are subject to change, objects in the heavens were created perfect and immutable. Rather than falling toward the Earth, the outer sphere(s) containing the celestial bodies moved in circular motion, eternally cycling back on itself with no beginning and no end.

The two-sphere model held sway for nearly two millennia, and as a result there was little progress in the physical understanding of how the Universe works. The perfection and constancy of the model, with celestial objects cycling along eternal circular tracks, made it appealing to Ptolemaic and later Christian sensibilities. But beyond the visual appeal of the heavens, the stars in the night sky were otherwise largely uninteresting. The stars, Moon, Sun and planets existed above the Earth, and the rules that govern change on the Earth were simply presumed to be invalid for the heavens. This meant that finding detailed explanations for astronomical phenomena wasn't particularly compelling to most people in the centuries leading up to the time of Copernicus (1473–1543), other than the impact that planetary positions had on keeping track of time, aiding in navigation, and the connection between human fate and the precepts of astrology. Questions were restricted to when astronomical phenomena would occur, and only because it was presumed that those phenomena impacted human affairs. Asking *why* the heavens move as they do, or ascertaining their origins, was not the province of educated philosophers who were concerned with the nature of humans rather than the nature of stars, planets and the Universe.

If the night sky were populated only with stars, then the two-sphere model might have continued for centuries longer than it did. However, then as today, planets captured the attention of early philosophers, scientists and tinkerers. While stars execute their uniform, circular motion about the Earth, the planets are the iconoclasts, breaking the rule of immutability. Venus and Mercury don't use the entire night sky as their playing field, but rather appear to the naked eye only close to when the Sun is setting or rising, swapping positions from one side of the Sun to the other. Venus in particular grabs people's attention to this day, appearing bright and bold just before and during sunset on some days, and around sunrise on other days, hence its designation as the evening star or morning star, respectively. Mars, Jupiter and Saturn are also extremely bright, yet they move across the entire night sky drifting across the background field of stars at their own paces.

Even more curiously, Mars, Jupiter and Saturn—along with the other outer planets, which are not visible to the naked eye—occasionally halt and then reverse course for weeks to months, from night to night moving from west to east, counter to their more typical east-to-west motion. From our vantage point of modern science, this *retrograde* motion, coupled with the restricted movement of Venus and Mercury, are evidence against an Earth-centered Solar System. Venus and Mercury never stray far from the Sun because the Earth is orbiting exterior to them and we are looking in toward their smaller, Sun-centered orbits. Meanwhile, Mars, Jupiter and Saturn orbit the Sun exterior to the Earth, so we can see them in their larger orbits that range over our entire night sky, often far from the Sun's position. Furthermore, the orbit of the Earth can "overtake" the orbits of the outer planets, causing them to appear to move backward on our sky as we pass them.

However, the prevailing way of thinking about the Universe before the sixteenth century compelled people to double down on the preference for perfect circular motion. Ptolemy built on the concepts of circular motion that were originally proposed by Hipparchus and Apollonius of Perga. Instead of traveling solely along giant circles centered on the Earth, the planets also moved along smaller circles centered on a point along the larger orbital circle, known as *epicycles*. This circle-on-a-circle concept allowed the planets to move from east to west on the sky most of the time, but the rotation of the epicycle allowed them to occasionally reverse direction. The epicyclic modification of Ptolemy was the dominant model of the Solar System for more than a millennium, from the second century A.D. up until the time of Copernicus in the sixteenth century.

1.3 Asking Why the Planets Move as They Do

For much of his life, Nicolaus Copernicus was employed variously as a politician, theologian, and physician. However, his true passion was astronomy and he was well-versed in the astronomical philosophy of Ptolemy, with its spheres and epicycles. However, the use of epicycles required not just small circles atop larger circles, but the model required other modifications such as offsets between the Earth and the center of the different planetary motions to reproduce, for example, the varying brightnesses and speeds of the planets throughout the year.

Copernicus found that circular planetary orbits could be accommodated by a Sun-centered (heliocentric) model (Sobel, 2011). Copernicus shared the Greek admiration for circles because an object on a circular path has eternal motion, which reflects the eternal perfection of the heavens. As an additional benefit, the arbitrary mechanism of epicycles was no longer needed to explain retrograde motion. He first proposed his heliocentric model in a short treatise entitled *Commentariolus* ("short commentary"). In this short work he laid out a set of assumptions, notably that there was no single center in the Solar System, but instead several rotation points; the Moon orbits the Earth; the objects other than the Moon orbit the Sun; the stars are fixed and at a great distance from the Solar System; and the apparent retrograde motion of the outer planets is related to the Earth's motion on a sphere interior to those planets' spheres. These assumptions formed the basis for his later work, *De Revolutionibus Orbium Coelestium* (On the Revolutions of the Heavenly Spheres), which described his heliocentric model more fully.

It should be noted that Copernicus' motivations for a heliocentric Solar System weren't particularly compelling as viewed from a modern scientific standpoint. Because of his insistence on circular motion, which we now know is not generally true of planetary motion, he proposed placing the Sun near the center of his system based on another aesthetic motivation: light tends to emanate from a central location, like light from a candle in an otherwise dark room. So he proposed to place the Sun at the center of the Solar System—the center of the room, so to speak— with the planets, including the Earth, revolving around the central light. Copernicus' reasoning for a heliocentric universe provided a simpler explanation than an Earth-centered model with epicycles. But more important, the heliocentric model later found overwhelming and compelling support in direct observational evidence.

His motivations aside, Copernicus' idea proved revolutionary because it provided an entirely new way of
observing and interpreting the Universe. Kuhn astutely
notes that the Copernican Revolution ushered in a new
paradigm of scientific thinking.[2]

A scientific paradigm is a system of thought that forms
a starting point for working out scientific problems. For
example, before the modern germ theory of disease, illness
was thought to be due to poor air quality, or "miasma."
Today, our understanding of bacteria, parasites and viruses
provides a much more effective framework for treating
and preventing diseases that would be unrecognizable to
medical practitioners of, say, the eighteenth century.

Similarly, the assumption that the Earth does not
occupy a special place in the Universe provides a common
starting point for working out modern astronomy problems, and this approach was largely unheard of before
Copernicus' revolutionary idea. Because of Copernicus'
paradigm shift the assumptions under which scientists
interpreted the world around them were fundamentally
changed. With the Sun moved to the center of the planets'
orbits, the Earth was no longer the center of the entire
Universe. With the new model, there was at last a system
of planetary motion around the Sun; there was now a
Solar *System*, a logical, universal mechanism for explaining
the motion of planets. Later observations and reasoning
revealed the Sun to be one of myriad stars in a Universe

[2]Owen Gingerich once told me the story of Thomas Kuhn, who extended
the definition of paradigm to the scientific realm. Thomas was one day walking
near Harvard Square, whereupon he was asked by a beggar if he could 'pare a
dime. I love a good pun!

that was much larger than just the Earth and its immediate environs.

Contrary to popular belief, the church initially took a rather pragmatic stance toward Copernicus' new idea. His heliocentric model provided a much easier method of computing predictions for the positions of the planets, and was therefore quite a bit more useful than previous astrometric methods. Copernicus was also a part of the church hierarchy, and he was very careful to follow proper rules and etiquette when publishing his ideas, going so far as to dedicate *De Revolutionibus* to the pope (Pogge, 2005). Immediately following the publication of *Commentariolus*, some religious leaders, both Catholic and Protestant, expressed disdain at the removal of the Earth from the center of God's creation. But there was not an organized, church-led suppression of Copernican thinking until several decades later. Copernicus was nonetheless sensitive to potential religious objections to his theory, which is one reason he waited until late in his life to publish *De Revolutionibus.*

Another obstacle to the widespread acceptance of the heliocentric model is the requirement of a moving Earth. A static Earth at the center of the Solar System fit into the existing philosophical beliefs that the elements associated with the Earth tend to fall toward its center. However, this tendency is distinct from the modern notion of gravity, and many scoffed at the notion of a moving Earth because it was felt that things resting on the Earth's surface, as well as the atmosphere and the Moon, would fly away if the Earth moved around the Sun. After all, if a horse-drawn cart laden with supplies turns a corner too quickly, the cargo will fly over the edge. Shouldn't the same hold true for the Earth and its "cargo"? Another objection was related to the firmly held belief that there could be only one center of motion, and having the Earth orbit the Sun as one center while serving as the center of the Moon's motion seemed absurd.

Copernicus' heliocentric model provided an adequate match to existing observations, but so did the prevailing Earth-centred (geocentric) model. To Copernicus, moving the Sun to the center of the planetary orbits provided a simpler, more mathematically elegant construct, but the tension between new and old theories would only be settled using the force of new observational evidence. If a model says astronomical objects—or the entire Universe, for that matter—should behave in a certain way, there are usually consequences that can be observed if that model is true. New observations of the Universe often provide tests of key aspects of theoretical models. Should the new theory pass the observational tests, then it can be adopted as a viable means of understanding the Universe, especially if the competing model cannot pass the same tests. Sadly, Copernicus died within a year of his theory's publication and therefore never saw the tests that led to the widespread acceptance of his hypothesis.

At the time of Copernicus, astronomical instruments were no more than basic surveyor's tools—means of measuring positions, angles and times in order to chart the motion of objects visible to the naked eye. Tycho Brahe (1546–1601) combined the latest versions of these tools with meticulous attention to detail to make marked improvements to past astrometric measurements of the planets. While Tycho strongly opposed Copernicus' heliocentric model, his collaborator Johannes Kepler inherited—or stole, depending on one's perspective—Tycho's data after he died, and used the highly precise data to refine Copernicus' model. Rather than uniform circular motion around the Sun, Kepler found that a modified law of motion could account for all aspects of planetary motion as manifest in Tycho's astrometric data.

The first of Kepler's laws states that planets move along ellipses rather than circular paths, with the Sun at one of the two foci. The second states that the closer a planet is to the Sun along its orbital path, the faster it moves, and vice versa, such that equal areas inside of each orbital ellipse are swept out in equal intervals of time. Today, we understand this through the lens of the conservation of both energy and angular momentum. The smaller the distance d between a planet and the Sun, at *perihelion*, the more negative the potential energy: gravitational potential energy goes as $-1/d^2$. In order to reduce its potential energy (note the negative sign) while conserving its total energy, the planet must increase its kinetic energy by speeding up. The opposite happens at *aphelion*, or the maximum orbital separation along the ellipse.

Kepler's third law is perhaps most familiar to physics students. It states that size of a planet's orbit, given by its semimajor axis a, is proportional to its orbital period, P, such that

$$(1.1) \qquad\qquad P^2 \propto a^3$$

Kepler's third law provided a unified mathematical framework for understanding the motion of the planets within

Copernicus' heliocentric model. The Universe containing the Earth, Sun, Moon and five naked-eye-visible planets was no longer an ad hoc collection of mechanisms, but rather a self-consistent system. However, lingering concerns about a moving Earth remained until the Solar System bodies could be inspected much more closely than was possible with the naked eye.

In the early 1600s, Dutch glass workers developed the first telescopes, which used systems of lenses to improve upon the eye's light-gathering ability and spatial resolution. These original telescopes were used for, e.g., spotting distant ships out at sea. A few years after the construction of the first telescopes, Galileo Galilei, an Italian polymath cut from the same cloth as Copernicus, started building his own telescopes and improved on the original Dutch designs. Rather than looking horizontally at objects on the Earth's surface, Galileo turned his improved telescope upward to the night sky. What he found from his newfound vantage point would have astounded the ancient Greek philosophers and surely warmed Copernicus' heart had he been alive to share the view.

Galileo's observations of the brightest object in the night sky, the Moon, revealed a crumpled, cratered landscape that stood in stark contrast to the supposed perfectly reflecting surface of a body made of ether, a perfect substance through which heavenly bodies moved and were constructed. The dark patches, rather than reflecting the imperfections of the Earth, were intrinsic to the Moon. Similarly, the projected image of the Sun's surface showed dark patches (sunspots) that changed in both size and location on timescales of days. Neither the Sun nor the Moon was perfect and immutable.

His close inspections of the planets were equally surprising. Venus was observed to exhibit phases during its orbit, similar to the Moon throughout the month. This showed convincing evidence for at least one planet orbiting the Sun rather than the Earth. The resolved image of Saturn showed that it was not perfectly circular, but rather had elongations on either side that change throughout the year. These elongations are known to be Saturn's characteristic ring system, but their variability was another knock against the unchangeable nature of the heavenly bodies.

The surprises kept on coming when Galileo examined Jupiter. Unlike Saturn, Jupiter resolved into a seemingly solid disk with no rings. But Galileo had a finely tuned attention to detail, and he noticed several "stars" that lay in a straight line that passes through Jupiter's center and always appear near Jupiter on the sky. Night after night, these attendant points of light would move from side to side and appear to change in number as one or two would disappear only to show up again in a new location the following night. Remarkably, from this apparently random jumping about on a two-dimensional surface, and despite interruptions due to poor weather, Galileo was able to infer that the points of light were executing three-dimensional orbital motion around Jupiter. This was profound because Jupiter suddenly became another center in the Universe, around which other bodies orbited. And since Jupiter orbits the Sun, it was evidence that the Earth could move through the Solar System without losing its Moon, atmosphere and human inhabitants.

1.4 Exoplanets and Completing the Copernican Revolution

More than four centuries after Copernicus' revolutionary idea we now live in an era of astronomy marked by another paradigm shift thanks to the discovery of planets around stars other than the Sun. The study of exoplanets—from their initial discovery to their physical characterization—is known as exoplanetary science.[3] The new knowledge flowing from this rapidly growing field is revolutionizing how we think of the origins of the Solar System specifically, and the formation and evolution of planetary systems throughout the Galaxy more generally.

Before the discovery of exoplanets our concepts of planetary systems were shaped and guided by a sample of one: our own Solar System. The downside of this sample-of-one approach should be fairly obvious. Focusing on our Solar System alone is like doing sociology based on solely investigating one's own life history. An approach to understanding humans by considering only oneself may be preferred by most children, but eventually we all grow up and realize that we cannot understand ourselves and our society without studying other humans. Similarly, we cannot hope to understand our own planet's origins, and the formation of planetary systems in general, without a large sample of planets other than those in the Solar System.

---

[3] Planets orbiting other stars were initially called extrasolar planets. However, Virginia Trimble noted that she had never heard of a planet residing *inside* of the Sun. While some might blanch at the mash-up of Greek and Latin, the more commonly used word today is "exoplanet."

The dominant view of planets that existed before the discovery of exoplanets is worth examining. Black (1995) provided a comprehensive yet pessimistic review of planetary science in an article entitled "Continuing the Copernican Revolution." Some of the noteworthy features of the Solar System that informed the dominant paradigm circa 1995 include:

1. The planets all lie more or less in the same orbital plane, with mutual inclinations less than 4 degrees. The exception is the innermost planet, Mercury.
2. The average angular momentum vector of the planets is nearly parallel to that of the spin angular momentum of the Sun.
3. The planets have nearly circular orbits, again with the exception of Mercury.
4. All of the planets orbit in the same direction, and with the exceptions of Venus and Uranus, they all spin in the same direction as their orbital motion and in the same direction as the Sun's spin.

Many of these features were noted hundreds of years earlier, around 1755, by Immanuel Kant. Based on the features of the Solar System, he proposed that the planets all trace a common origin to a flattened, rotating distribution of gas—a nebula.[4] This *nebular hypothesis* was expanded upon by Pierre-Simon Laplace, who envisioned

---

[4]Kant also correctly reasoned that the Milky Way galaxy is a flattened disklike distribution of stars orbiting a common center.

the nebula cooling and contracting to form successively larger rings moving away from the newly formed central Sun. These rings then collapsed to form the planets.

While the nebular hypothesis of Laplace and Kant ran into speed bumps between the eighteenth century and mid-twentieth century, by the time of Black's review article the notion of planets forming out of a flattened distribution of gas and dust surrounding a newly formed star was the dominant model of planet formation. Indeed, it still is. We think that the process of planet formation is a by-product of the process of star formation. In the interstellar medium—the collection of gas and dust that roams in the vast expanses of the Galaxy between stars—there exist large clouds containing relatively dense, cool pockets of molecular gas. These molecular clouds experience compression, perhaps due to the shock fronts emanating from supernovae explosions from nearby star formation regions, and as a result start to contract under the pull of their own self-gravity. If the force of gravity cannot be counterbalanced by the internal thermal support of the molecular cloud, the cloud begins to collapse in on itself.

Molecular cloud collapse leads to one or many central concentrations that become stars. As the molecular cloud material condenses toward the central star, the conservation of angular momentum leads to the formation of a flattened, spinning disk containing gas and dust; material can compress along the spin axis of the natal star/disk. Turbulence, viscosity and magnetic fields help transport the inflowing gas through the disk and onto the central star, while stellar winds and irradiation push disk material back out into the interstellar medium. The leftover dregs of the star-formation process go into forming larger and larger collections of solid material that form the seeds of planet formation.

The planet formation process is brief compared to the lifetime of the central star. Stars like the Sun have lifetimes of roughly 10 billion years. However, the planet-formation epoch lasts only 10 million to 100 million years, or a period several orders of magnitude shorter than the life of the eventual system containing planets and the host star. Before the diversity of exoplanetary systems was observed, the Solar System was generally assumed to represent a static "fossil record" of the planet-formation process. Small planets formed close to the star, where the density of gas was smaller than in the outer regions, and where the temperatures were too hot for the condensation of volatiles such as water ice. However, beyond the so-called snow line, the disk temperatures dropped to the point at which ices could condense and solid planetary cores could form more easily. Together with the increased density of gas the giant planets were able to form more efficiently, hence Jupiter and the other gas giants at large orbital distances.

The discovery of exoplanets immediately challenged this picture of gas-disk-to-planet fossil preservation. Exoplanet architectures provide strong evidence of post-disk gravitational jostling among planets. For example, gas-giant planets around other stars are often discovered in orbits much smaller than the 5.2 astronomical units (AU) that separate the Sun and Jupiter. Also, gas-giant exoplanets are often in eccentric orbits, and sometimes the orbits are significantly tilted with respect to other planets in the system, and with respect to the stellar spin axis. The prevalence of eccentric, tilted, close-in planets has led to the invocation of gravitational interactions between planets and their disks, between planets and their stars, and among planets. With a sample of one, it would be difficult to make a strong argument for such radical gravitational interactions and inward orbital migration. But with the view afforded by thousands of alien planetary systems, we can now see that the orderly architecture of the Solar System may be an exception rather than the rule throughout the Galaxy.

An analogy for our view of planetary systems before and after the discovery of exoplanets is the story of a Victorian house in the middle of an urban area such as Manhattan.[5] Imagine growing up in such a house, but never having the opportunity to look outside, and therefore never seeing other dwellings where other people live. Then, one day at the age of twelve you are afforded the opportunity to step outside and look around. The high-rise apartment complexes, townhouses and condominiums would look nothing like the house you spent your whole life in. Any theories of how other people live would be radically changed, and instantaneously so. At the same time, your mind would be flooded with basic but perplexing questions such as, How do people get up into their dwellings from the ground level? How were these towering complexes built and who built them? Do people collectively own the entire building, or do they own only parts of the building?

---

[5] Here, I'm envisioning the house in one of my favorite children's books, *The Little House*, by Virginia Lee Burton.

All of these basic questions have fairly basic answers, but those answers would be far from obvious on the first few days outside of the little Victorian house. Similarly, the large number and variety of planetary systems that we have found elsewhere in the Galaxy are causing us to confront similarly basic yet perplexing questions. Having "grown up" in a Solar System with a system of small, rocky planets close to the Sun (Mercury, Venus, Earth and Mars), and a system of gas giants further out (Jupiter, Saturn, Uranus and Neptune), how do we understand the existence of hot Jupiters: gas-giant planets orbiting right next to their host stars with orbital periods of days rather than years (Schilling, 1996)? How do we explain the formation of planetary systems containing three to seven sub-Neptune-sized planets all packed into regions smaller than the Sun–Mercury separation (Mayor et al., 2009; Lissauer et al., 2011; Swift et al., 2013)? What context do we have for Saturn-sized planets that orbit not one, but two stars—circumbinary planets from which one would see vistas similar to that of Luke Skywalker standing on Tatooine, the fictional double-Sun planet in the movie *Star Wars* (Doyle et al., 2011; Orosz et al., 2012)?

Our models of planet formation, which had for decades been informed by and tuned to explain our own Solar System and its architecture, must now undergo radical revision. Describing the Solar System is no longer the only challenge facing planet-formation theorists. Indeed, as of only 2014 we now know that our Solar System is not even a typical planetary system in the Galaxy. Tiny red dwarf stars outnumber stars like the Sun by a ratio of about ten to one, and there are one to three planets per red dwarf (Dressing & Charbonneau, 2013; Morton & Swift, 2014). Just as with Copernicus' radical theory, in which the Earth was dethroned from its central position in the Solar System, thanks to the discovery of exoplanets the entire Solar System has been moved from its position of primacy in our thinking about planetary systems in general. The birth and rapid rise of exoplanetary science has fundamentally and radically changed our paradigm for understanding the existence, origin and evolution of planets on a galactic scale.

2

STELLAR WOBBLES

> But there seems to be no compelling reason why such hypothetical ... planets should not, in some instances, be much closer to their parent stars than is the case in the solar system. It would be of interest to test whether there are such objects.
>
> — *Otto Struve, 1952*

2.1 At the Telescope

"Carolyn, we're ready to move to the next target." I'm talking to the Keck Observatory observing assistant, who is at the controls of one of the world's largest telescopes, one of the twin 10-meter Keck telescopes. To be honest, it's not even really necessary for me to make this request to Carolyn since she knows the routine of the California Planet Survey as well as I do, if not better. If the weather is good, we will methodically march down our target list, observing approximately one star every five minutes, for a grand total of about 120 stars in a single night. Before I can even finish my sentence, the control screen in front of me indicates that the telescope is on its way to the celestial coordinates of the next target of the night.

Many older astronomers blanch at the way that modern observational astronomy is conducted. For example, we are sitting in an environmentally controlled room at sea level in Pasadena, California. We're working at a computer with three monitors that display every vital piece of information about the Keck 1 Telescope and the instrument we're using, the HIgh-Resolution Echelle Spectrometer (HIRES). We're connected to the telescope across the Pacific Ocean via a high-bandwidth fiber-optic cable that mediates my conversation with the observing assistant, Carolyn. Unlike me, she is actually at the telescope, atop an extinct volcano called Mauna Kea on the Big Island of Hawai'i.

This is a far cry from the days when astronomers would travel to the mountaintop, put on heated bomber jackets and climb up to the prime focus of a telescope such as the Palomar 200-inch near San Diego. There at the top of the telescope, 40 feet above the dome floor in a tiny space-capsule-like box, the intrepid astronomer would spend the night observing. Perched in the observing "cage" for 9 to 12 hours with no breaks, the astronomer's job was to keep the night's target stars centered on the crosshairs of an eyepiece while sending fine guiding controls to the telescope via a wired, push-button controller. These were long, painstaking nights spent alone in the cold and darkness, without the ability to take breaks to walk around or stretch. As George Herbig once told me, "You'd take two thermoses with you up to prime focus. One empty and one full. At the end of the night, you'd bring two thermoses down: one empty and one full." There were no restroom breaks up at prime focus.

Nowadays we have computers to guide the telescope, and fiber-optic cables to carry a multitude of information about the telescope to observers sitting in a warm room at sea level. Modern observational astronomers spend far less time and effort at the telescope than our earlier colleagues did. However, they spend just as much time planning their observing runs, analyzing their data and publishing their results. Computers rather than humans move the telescope across the sky, and the images and spectra are recorded on digital detectors rather than photographic plates. But the data still require a human interpretation, and the knowledge we glean from those data are just as critical as ever for our understanding of the Universe.

On this particular night, I am excited about star number 96 on our target list, also known as HD 94834, the 94,834th star in the Henry Draper Catalog of bright stars. Admittedly, it's a fairly nondescript star in the constellation Leo. It is too faint to warrant a Bayer or Flamsteed designation, like α Cen A or 70 Virginis. A search for the star in the astronomical literature shows that it appears in a catalog of stellar distances, and a catalog of stellar magnetic activity. However, to me this star is special because it affords an opportunity to see planets that were hidden from view during most of its prior lifetime. I started observing it back in the fall of 2007 as part of my search for planets around stars more massive than the Sun. HD 94834 reached the end of its main-sequence life about 10 million years ago—just yesterday in astronomical terms. On the main sequence, it "worked a nine-to-five job" generating the energy needed to hold itself up by fusing hydrogen in its core.

After about 4.6 billion years HD 94834 entered into "retirement" following its life of fusing hydrogen in its core on the main sequence. Having exhausted its source of hydrogen fuel, the core began to contract. Paradoxically, as the core contracted, the rest of the star expanded in order to maintain hydrostatic equilibrium, and as a result, the star cooled. More important to me, the star also dramatically slowed its rotation,[1] thereby bringing the absorption lines in its spectrum into better focus. As part of my planet search program, I monitor the velocities of my target stars looking for accelerations along our line of sight that belie the existence of previously unknown planets. As we will see later in this chapter, these accelerations are observed as tiny shifts in the absorption spectra of the star due to the gravitational tug of an unseen planet. While on the main sequence, HD 94834 was rotating two orders of magnitude more rapidly than it does today. This rapid rotation smeared out its absorption lines, thereby obscuring information about the star's planets. But in retirement, the star has become an optimal target for my planet-search program.

The first three velocity measurements I made of HD 94834 back in 2007 and 2008 were completely flat, and initially the star appeared to be moving at a constant velocity with respect to the Solar System. However, the fourth velocity measurement made in the winter of 2008 showed a deceleration as the star's velocity decreased by 35 m s^{-1}, at a speed where it appeared to remain for the next year.

---

[1] The star slowed in part because of the conservation of angular momentum as it expanded its radius. Another reason is that its magnetic field interacted with the stellar wind, allowing material to be carried along the magnetic field lines to large distances from the star. This transports angular momentum away from the star, allowing it to dramatically slow its rotation.

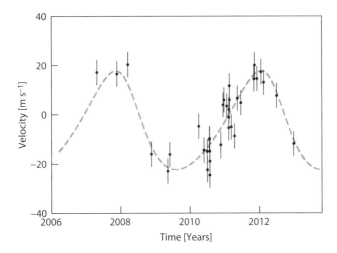

Figure 2.1. Radial velocity measurements of HD 94834 made at Keck Observatory with the HIRES spectrometer. The error bars represent the uncertainty due to instrumental errors and astrophysical "jitter" (see Section 2.5). The dashed line shows the best-fitting orbit model. Later in this chapter I demonstrate that the properties of this planet (minimum mass, period, eccentricity and semimajor axis) can be read by eye.

Then in 2010 the star began to noticeably accelerate once again, steadily changing its velocity by about 17 meters per second per year (m s^{-1} yr^{-1}) until it returned to its 2008 velocity. In 2012 I noticed that it began to decelerate once again, making it a candidate for exoplanet discovery. However, at the moment it is only a possible planet that awaits confirmation from additional observations. Tonight at the telescope I am hoping that the photons that hit our detector carry information about the candidate planet's orbit. If so, I will have discovered a new planet in a 4.3-year, slightly eccentric orbit around the subgiant star HD 94834. The planet will be named according to tradition: HD 94834 b. The backstory of how astronomers identify and measure the properties of distant planetary systems is tied to the physics of orbital motion, which is the subject of the following section.

2.2 For Every Action

From our view within our Solar System one could ignore the fact that planets orbit the Sun and one would still be able to detect them and study their physical structure. Indeed, for thousands of years humans studied the night sky's wandering bodies with little to no knowledge of the gravitational bond between the Solar System planets and the Sun. Planets were the bright, nontwinkling objects that moved differently throughout the year, in contrast to the predictable motion of the stars.

However, to find planets orbiting other suns, it is crucial to have a correct understanding of the proper relationship between stars and their planets. Planets like those in our Solar System are gravitationally bound to stars, and as they orbit they can affect the light emitted from stars in different ways, depending on the geometric configuration of the system with respect to the observer. Perhaps the most general effect is the planet's gravitational tug on its star.

Most people who have taken a basic physics course know that stars exert a gravitational tug on their planets causing them to stay in orbit. However, it is not strictly correct to picture the planet orbiting a stationary star. This is because, as Newton posited, every action has an equal and opposite reaction. As the star exerts a gravitational tug on its planet, keeping it in orbit, the planet tugs back on the star with a force that is equal and opposite of the star's gravitational pull. However, since the planet is so much less massive than the star, and because the force is equal to the mass of an object times its acceleration ($\vec{F}_{net} = m\vec{a}$), the resulting acceleration of the star is very small because its mass is very large.

In the following sections I derive the magnitude of a star's motion under the simplifying assumption of circular orbits, discuss the scaling relationships between the properties of the star's observed motion and the physical properties of the planetary system, and offer a brief overview of the methods used to measure the velocities of stars to high precision.

2.2.1 Radial Velocity Variations for a Circular Orbit

The famous seventeenth-century astronomer Johannes Kepler found that the planets in the Solar System move on elliptical (eccentric) orbits. However, it is instructive to first consider the special case of circular orbits (zero eccentricity, $e = 0$), since this case embodies most of the basic physics of orbital motion. Later we revisit the equations derived for circular motion and include the modifying effect of nonzero eccentricity.

Consider a snapshot of a star and its planet in a circular orbit, as viewed from above the system, i.e., pole-on (Figure 2.2). The star and planet will lie along a straight

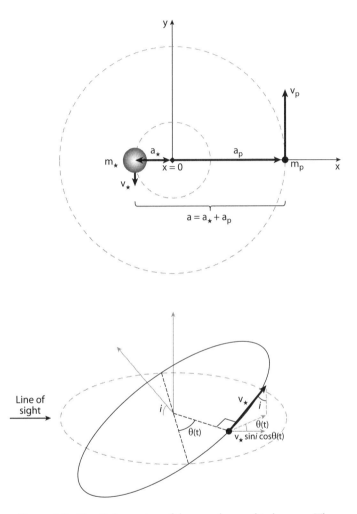

Figure 2.2. *Top:* Pole-on view of the star-planet orbital system. The star and planet orbit their mutual center of mass, or barycenter, which corresponds to $x = 0$, with the x-axis running from left to right. The planet's distance from the center of mass is given by $x = a_p$, while that of the star is $a_\star$. The mean semimajor axis is

line, with velocities perpendicular to this line. The point at which the opposing gravitational forces are balanced between the star and planet is the center of mass, or barycenter of the system. Rather than planets orbiting their star, the actual situation involves the star and its planets orbiting their mutual barycenter, which we'll assume to be at rest with respect to the observer.[2]

The center of mass of two objects is placed at the origin of the x-axis, as in Figure 2.2, and the objects have masses $M_\star$ and m_p, for the star and planet, respectively. If $a_\star$ and a_p denote the distances of the star and planet, respectively,

---

[2]The barycenter of exoplanetary systems are generally moving with respect to the Solar System and other stars, because stars in the Galaxy have random motions with respect to one another.

Figure 2.2. (*Continued*)

defined to be $a = a_\star + a_p$. The semimajor axes are related by $a_\star = (m_p/M_\star)a_p$ and the velocities are related by the mass ratio, such that $v_\star = (m_p/M_\star)v_p$. *Bottom:* An illustration of the star's orbit. Unlike in the top panel, we are viewing the star's circular orbit at an arbitrary orientation in space, with a nonzero inclination i along our line of sight. Pole-on orbits correspond to $i = 0$ degrees and edge-on orbits have $i = 90$ degrees. The radial component of the star's velocity is given by the dot product of the velocity vector $\vec{v}_\star$ and the unit vector pointing from the observer (at left) to the orbit's barycenter, with an observed radial velocity $v_{\rm rad} = v_\star \sin i \cos \theta(t)$. Positive velocities are away from the observer, and negative velocities correspond to motion toward the observer. The maximum velocity is reached at $\theta(t = 0) = 1$, corresponding to $K = v_\star \sin i$, which is the amplitude, K, of the RV signal caused by a planet with a minimum mass $m_p \sin i$. If $i = 0$ degrees, then $\sin i = 0$ and there is no component of the star's motion along the line of sight.

from the barycenter, we can make a choice of the origin and set the location of the barycenter to $\bar{x} \equiv 0$. The equation for the center of mass is given by

$$(2.1) \qquad \bar{x} = \frac{-M_\star a_\star + m_p a_p}{M_\star + m_p} = 0$$

where the sign on the first term in the numerator is negative because the star is at a negative distance from the origin, $x_\star = -a_\star < \bar{x}$, whereas the planet is at a positive distance $a_p > \bar{x}$. Note that in this context, $a_\star$ and a_p are positions, with units of distance, and should not be confused with acceleration, which is denoted by the same lowercase a symbol. Additionally, we define the "mean semimajor axis," $a = a_\star + a_p$. This quantity will reappear further along when we derive Newton's version of Kepler's Third Law.

Solving Equation 2.1 for the star's semimajor axis gives

$$(2.2) \qquad a_\star = \frac{m_p}{M_\star} a_p$$

which shows that the star's distance from the center of mass is proportional to the planet's distance times the planet-star mass ratio. Since the planet's mass is so much less than the star's mass ($m_p \ll M_\star$), $a_\star$ will be much smaller than a_p.

The relationship between a_p and $a_\star$ provides a relationship between the speeds of the planet and star. Note that the star and planet traverse the circumference of their orbits, $2\pi a_\star$ and $2\pi a_p$, respectively, once every orbital period P. Since speed is the distance traveled divided by the time needed for the trip, this leads to

$$(2.3) \quad v_\star = \frac{2\pi a_\star}{P} = \frac{2\pi \left(\frac{m_p}{M_\star} a_p \right)}{P} = \frac{m_p}{M_\star} v_p$$

This shows that the speeds of the star and planet are also related by the mass ratio $m_p / M_\star$. Because the ratio $m_p / M_\star$ is very small, the speed of the star will be much less than the speed of the planet. This should agree with your physical intuition: the same force acting on a massive body (the star) will accelerate it less than if the force acted on a less massive body (the planet). Also note that $m_p v_p = M_\star v_\star$, which shows that the magnitude of the star and planet momenta are equal. This is to be expected when viewing the orbital system from the perspective of the center of mass: momentum is conserved.

Having expressions for the speeds of the star and planet is useful because a star–planet gravitational system can be considered a "virialized system." As long as the system doesn't change its shape (moment of inertia) over time baselines that are long compared to the orbital period, it can be adequately described by the average kinetic and potential energies of its constituents. Kinetic energy involves an object's speed and mass of the gravitationally bound objects, and the potential energy is related to masses and separation of the two objects. An orbit satisfies the conditions of a virialized system, since the time-averaged shape of a two-body orbit is constant (the same holds for stable multibody orbits). Working with the energy of the system allows us to find a relationship among the period, semimajor axis and stellar mass. Of the many ways to derive this relationship, I find the following derivation highly intuitive and simple, and familiarity with the virial theorem is useful for solving a wide range of astrophysical problems.

The virial theorem states that at all times the kinetic energy, $\mathcal{K}$, of a virialized system is half the potential energy holding the system together, U, or

$$\mathcal{K} = -\frac{1}{2}U$$

$$(2.4) \qquad \frac{1}{2}m_p v_p^2 = -\frac{1}{2}\left(-\frac{Gm_p M_\star}{a}\right)$$

where I have replaced a_p with a since $m_p \ll M_\star$ and $a_p \gg a_\star$. Solving for v_p gives

$$(2.5) \qquad v_p = \sqrt{\frac{GM_\star}{a}}$$

Note that this same expression can be derived by setting the centripetal acceleration v_p^2/a, equal to the gravitational force per unit mass, $GM_\star/a^2$.

Substituting $v_p = (M_\star/m_p)v_\star$, and $v_\star = 2\pi a_\star/P$ in Equation 2.5 and doing a bit of algebraic manipulation results in

$$(2.6) \qquad P^2 = \frac{4\pi^2 a^3}{GM_\star}$$

This relationship is equivalent to Newton's version of Kepler's Third Law of planetary motion under the assumption that the star's mass is much larger than the planet's mass ($M_\star + m_p \approx M_\star$). If the period, P, is measured in years, the semimajor axis a is in astronomical units (AU; where 1 AU is the Earth–Sun distance) and the star's mass is equal to that of the Sun ($M_\star = 1\ M_\odot$), then we can recover Kepler's original Third Law,

$$(2.7) \qquad\qquad P^2 \propto a^3$$

We now have all of the puzzle pieces necessary to understand the motion of the star in response to an orbiting planet. We have figured out how the speeds of the star and planet are related (Equation 2.3), and we have figured out the relationship among the period, semimajor axis and mass of the system, with the mass dominated by the central star $M_\star \gg m_p$ and $a_p = a$ (Equation 2.6).

Now let's put it all together starting with the expression of the planet's orbital speed in Equation 2.5. The planet's speed is related to the star's speed via the mass ratio, giving

$$(2.8) \qquad v_\star \left(\frac{M_\star}{m_p} \right) = \left(\frac{G M_\star}{a} \right)^{1/2}$$

Since the semimajor axis, a, is not an observable quantity with the radial velocity technique, we can use Kepler's Third Law to replace a with the planet's orbital period, P, and the stellar mass, $M_\star$. After a bit of algebra, the star's speed is given by the simple product of four quantities, one comprising various constants and the other three related to the physical properties of the system:

$$(2.9) \qquad v_\star = (2\pi G)^{1/3} M_\star^{-2/3} P^{-1/3} m_p$$

If we express the stellar mass, period and planet mass in terms of the mass of the Sun ($M_\odot$), the period in years, and the planet mass in Jupiter masses (M_{Jup}), respectively, we can recover the useful, quantitative relationship

$$v_\star = [28.4 \text{ m s}^{-1}] \left(\frac{M_\star}{M_\odot} \right)^{-2/3}$$

$$(2.10) \qquad \times \left(\frac{P}{\text{year}} \right)^{-1/3} \left(\frac{m_p}{M_{\text{Jup}}} \right)$$

Thus, a Jupiter-mass planet in a one-year orbit around a solar-mass star will cause its star to move at a speed of 28.4 m s^{-1}. To put this in perspective, 1 meter per second is about 2.2 miles per hour. A Jupiter-sized planet in a one-year orbit would cause the star to move about 62.5 miles per hour. This is decently fast for a car, but it is extremely slow by astrophysical standards. For example, the Sun orbits the center of the Milky Way Galaxy at a speed of 486,000 miles per hour (roughly 220 km s^{-1}), which is typical for stars in the Galaxy. The velocity variations induced by planets are a tiny factor on the movement of a typical star.

We can also express the star's speed in terms of the more physical variable a, the semimajor axis, rather than the period P. This can be done using Kepler's Third Law to replace P, resulting in

$$v_\star = [28.4 \text{ m s}^{-1}] \left(\frac{M_\star}{M_\odot} \right)^{-1/2}$$

$$(2.11) \qquad \times \left(\frac{a}{1 \text{ AU}} \right)^{-1/2} \left(\frac{m_p}{M_{\text{Jup}}} \right)$$

If we instead express m_p in Earth-masses, the numerical factor in front becomes much smaller. Jupiter is about 318 times more massive than the Earth, so $v_\star$ becomes 318 times smaller, or 9 cm s^{-1}. Stop and think for a moment about this speed: 9 *centimeters* per second. The width of a piece of printer paper is about 20 centimeters. Trace your finger across the sheet such that it takes two seconds to go from one side to the other. This is how fast the Sun—a giant ball of gas, 100 times the diameter of the Earth—moves in response to the Earth's gravitational tug. And somehow astronomers must detect the motion of this ball of gas moving at 10 centimeters per second from a distance of tens or hundreds of light-years in order to find Earth-mass exoplanets at 1 AU around their stars.

Throughout this derivation we have assumed that we are viewing the planet's orbit edge-on, so the line-of-sight component of the stellar velocity, its radial velocity $v_{\rm rad}$, is at its largest value. However, imagine a situation in which the orbit plane is parallel to the sky such that we are viewing the planetary system from "above," or pole-on, rather than edge-on. In this case, the planet and star will always move perpendicular to our line of sight, and we would detect zero motion in the radial direction. This is because we see only the projection of the star's velocity vector along our line of sight, and a face-on orbit has no portion of the star's motion aimed toward or away from us.

If the star orbited by a Jupiter-mass planet with a period of 1 year were observed from afar with an edge-on orientation, the maximum velocity along the line of sight, given by $K \equiv \max(v_\star)$, would be 28.4 m s^{-1}. The observable, K, is often referred to as the radial velocity "semi-amplitude" of the orbit. As can be seen in Equation 2.10, K is directly proportional to the planet's mass: $K \propto m_p$. However, since we do not typically know the inclination of the orbit, the mass of the planet is degenerate with $\sin i$, and the value of m_p inferred from the amplitude represents the planet's *minimum mass*, expressed as $m_p \sin i$. If the inclination is smaller, then $\sin i$ is smaller, and the planet's true mass is larger than one would measure from K.

Next, we must account for the time variability of the star's radial velocity. The previous expressions give the star's *speed*, or equivalently the amplitude of the time-variable motion along the line of sight. But for a circular orbit the line-of-sight component of the star's velocity will vary along the orbit as a cosine wave of the form

$$(2.12) \qquad v_{\text{rad}}(t) = K \cos\left[\theta(t) - \omega\right]$$

where

$$(2.13) \qquad \theta(t) = \frac{2\pi\left(t - T_p\right)}{P}$$

I have added two important terms that relate to the phase of the planet's orbit. The first is ω, the *argument of periastron*, which describes the rotation of the orbit with respect to the observer. For an eccentric orbit, periastron corresponds to the planet's closest approach to the star, while apastron is the maximum distance between the star and planet.[3] For a circular orbit there is no actual periastron, so ω can be arbitrarily set to zero, which is

---

[3] As a mnemonic device, I think of "peri" as being my friend, who I want close. Apastron starts with "a," like "away."

defined as the point along the orbit when the angle $\omega(t)$ is such that the line connecting the star and the barycenter is perpendicular to the line of sight, with the star moving away from the observer, as shown in Figure 2.2. As we'll see later, ω can be nonzero for eccentric orbits, since an elongated orbit can be oriented with an arbitrary angle with respect to the observer's line of sight.

The second newly introduced term is the time of periastron passage, T_p. This term sets the phase of the orbit. Even though we have defined $\omega = 0$ for a circular orbit, the observed sinusoidal motion of the star's radial velocity will be random—different orbital systems will reach their maximum radial velocities at different times. For this reason T_P corresponds to the time when the planet passes through periastron.

The time between consecutive peaks gives the orbital period P. When combined with the amplitude of the signal, K, Equation 2.10 can be used to measure the minimum mass of the planet, $m_p \sin i$, divided by the mass of the star raised to the 2/3 power, or $m_p M_{\star}^{-2/3}$. Thus, in order to estimate the planet's minimum mass, one needs to know the star's mass. This is a theme that is repeated throughout this book: to understand the physical properties of planets, one must understand the physical properties of the stars they orbit.

2.3 Eccentric Orbits

To keep things fairly intuitive and easy to follow, up to this point I have neglected the effects of eccentricity throughout the derivation of the radial velocity of a star with a planet. Doing this is fine for getting a handle on the basic scaling of the problem, but as Kepler's First Law states, planets move in elliptical orbits. And while most of the planets of the Solar System have nearly circular orbits with $e < 0.1$—with the exception of Mercury, which has $e = 0.2056$—exoplanets have been discovered with a wide range of eccentricities varying from nearly circular up to cometlike orbits with $e = 0.934$ (HD 80606 b; Naef et al. 2001).

Including orbital eccentricity, e, also modifies the equation for the Doppler amplitude, K, through a multiplicative term involving e:

$$K = 28.4 \text{ m s}^{-1} \left(\frac{M_\star}{M_\odot} \right)^{-2/3} \left(\frac{P}{\text{year}} \right)^{-1/3}$$

$$(2.14) \quad \times \left(\frac{m_p \sin i}{M_{\text{Jup}}} \right) (1 - e^2)^{-1/2}$$

The time variability of the radial velocity of a star, v_{rad}, is given by

$$(2.15) \quad v_{\text{rad}}(t) = K(\cos[\theta(t) - \omega] + e \cos \omega)$$

However, in an eccentric orbit $\theta(t)$ does not vary linearly with time. Instead, $\theta(t)$ is related to a new variable called the *eccentric anomaly*, $E(t)$, by

$$(2.16) \quad \tan \frac{\theta(t)}{2} = \left(\frac{1 + e}{1 - e} \right)^{1/2} \tan \frac{E(t)}{2}$$

When $e = 0$, Equation 2.16 reduces to $\cos E(t) = \cos \theta(t)$, which is equivalent to $E(t) = \theta(t)$, where $\theta(t)$ is given by Equation 2.13 and includes the time of periastron passage, T_p, and the argument of periastron, ω.

The eccentric anomaly can be computed at a time t and an eccentricity e through the transcendental equation

$$(2.17) \quad E(t) = e \sin E(t) + \frac{2\pi (t - T_p)}{P}$$

At this point, why we did the derivations for the much simpler case of a circular orbit should be clear. Whereas a proper derivation requires consideration of the eccentricity, its derivation involves more geometry and algebra than physical insight. In practice, solving for the velocity of a star as a function of time can be accomplished with the following algorithm:

- Start with an initial time, t_0, in Equation 2.17 and numerically solve for $E(t_0)$ using an iterative procedure, such as a Newton-Raphson scheme.
- Solve for $\theta(t_0)$ by plugging e and $E(t_0)$ into Equation 2.16.
- Use $\theta(t_0)$ to compute $v_{rad}(t_0)$ using Equation 2.15.
- Repeat for a new time, $t_1 = t_0 + \Delta t$, where Δt is a time increment that is small compared to the orbital period.

The need to solve a transcendental equation makes computing Keplerian orbits fairly slow. Every ten years or so someone devises an improved method of solving Kepler's equations, which allows for more efficient

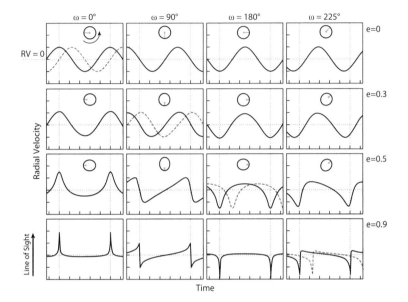

Figure 2.3. Illustration of the effects on the radial velocity of a star with varying the eccentricity e and argument of pericenter ω. The small orbit at the top of each panel shows the orbital motion of the *planet* with a period P (arbitrary units and magnitude), eccentricty e, orientation ω and phase T_p. The gray line in each orbit configuration indicates the angle ω with respect to the line of sight to the system (the angle ω is also labeled for each column along the top of the figure). The solid line showing the RV variation

simulations of N-body gravitational problems (Wisdom & Holman, 1991), and faster fitting of radial velocity time series through partially linearizing the equations (e.g., Wright & Howard, 2009) or through novel use of computing hardware (Ford, 2009).

Example orbits with different eccentricities, e, and phases (ω and T_P) are illustrated in Figure 2.3. In each panel, the planet's orbit, shown pole-on, is illustrated, with the observer assumed to be viewing the system from a vantage point at the bottom of the page. From this vantage point, the hypothetical observer will see the orbit edge-on. The vertical dotted line on the left in each panel denotes $t = 0$. Different values of ω rotate the orbit with respect to the observer.

Note that the corresponding radial velocity plots in each panel correspond to the star, not the planet. The planet and the star move counterclockwise in each panel. The diagonal panels, moving from upper left to lower right, illustrate the effect of changing T_p by a quarter of the orbital period. This has the effect of shifting the radial velocities to the right.

A great deal of information is contained in this figure, and understanding all aspects of it will not be achieved immediately. However, taking time to think through each example will help the student gain a physical intuition for reading radial velocity time series.

Figure 2.3. (*Continued*)
in each panel corresponds to the radial motion of the *star* (not the planet, which is shown above). Note that the stellar orbit that gives rise to the RV signal has a sign opposite to that of the planetary motion, corresponding to a phase difference of π. The times $t = 0$ and $t = P$ are indicated by the vertical, dotted lines. The gray, dashed curve is the same RV signal with the time of periastron changed to $T_P = P/4$, demonstrating how T_P governs the phase of the periodic RV variations. The rows show orbits with constant e as indicated on the right of the figure, and the columns are for constant ω as indicated along the top of the figure. In the first three rows, the orientation and shape of the orbit of the planet around the central star (gray circle, not to scale) are shown, with the orbit viewed from the bottom of the page.

2.3.1 Example: Reading Radial Velocities

Figure 2.1 shows the radial velocity (RV) measurements of HD 94834 made with the HIRES instrument at the Keck Observatory, along with the best-fitting orbit model shown as a dashed line. We can estimate the period by measuring the distance between consecutive peaks, which gives $P \approx 4.3$ year. The Doppler amplitude K is given by the distance from RV = 0 to the maximum value, or $K \approx 20$ m s^{-1}. Making the assumption of a circular orbit, and using the approximate stellar mass $M_\star = 1.3$ M$_\odot$, we can estimate the minimum planet mass by solving Equation 2.10 for $m_p \sin i$:

$$\frac{m_p \sin i}{M_{\mathrm{Jup}}} = \left(\frac{K}{28.4 \text{ m s}^{-1}}\right)\left(\frac{M_\star}{M_\odot}\right)^{2/3}\left(\frac{P}{\text{year}}\right)^{1/3}$$

$$(2.18) \quad m_p \sin i = \left(\frac{20}{28.4}\right)(1.3)^{2/3}(4.3)^{1/3}$$

$$\approx 1.4 \; M_{\mathrm{Jup}}$$

Using Kepler's Third Law of motion (Equation 2.6) we can also solve for the planet's semimajor axis:

$$a = \left(\frac{G M_\star}{4\pi^2}\right)^{1/3} P^{2/3}$$

$$(2.19) \quad = [1 \text{ AU}] \left(\frac{M_\star}{M_\odot}\right)^{1/3}\left(\frac{P}{1 \text{ year}}\right)^{2/3}$$

$$= (1.3)^{1/3}(4.3)^{2/3} \text{ AU} \approx 2.9 \text{ AU}$$

2.4 Measuring Precise Radial Velocities

Astronomers measure the velocity of stars using the Doppler shift induced in a star's light as it moves. We encounter the Doppler effect in our daily lives as a car approaches, passes, and then drives away from us. At first the engine sound is high pitched, and it gradually sounds lower pitched as it passes and then drives away from us. This is because the sound waves are compressed as the car approaches, and then stretched as it recedes. Since light can be described as a wave, the Doppler effect will cause light waves to become compressed ("blueshifted") as a star moves toward the Earth, and stretched ("redshifted") as the star moves away.

Despite how it is usually depicted in widely used illustrations, the Doppler technique of finding planets does not involve first looking for blue light, and then looking for red light as the star moves in different directions. At all times the distribution of energy emitted from a star is to first-order a blackbody, or Planck distribution, dependent only on the star's surface temperature. However, inspection of a star's spectrum shows deviations from that of a perfect, smooth blackbody emitter. These deviations are due to the absorption and emission of photons by atoms and molecules in the star's atmosphere, and the positions of these "lines" encode information about the star's motion.

The centers of stellar absorption lines correspond to the quantum levels of atoms and molecules in the star's atmosphere, and photons on their way out of the star encounter various atoms and molecules. If the wavelength of a photon corresponds to the energy difference between

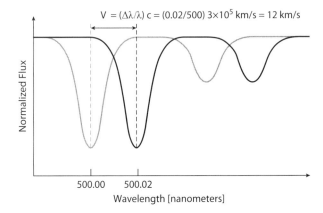

Figure 2.4. Illustration of a stellar spectrum at rest (gray) and the same spectrum Doppler-shifted (black). The change in wavelength between the two spectra is $\Delta\lambda = 0.02$ nm, and the center of the line was originally located at $\lambda = 500$ nm. The ratio of the change in the line's position to its original position, when multiplied by the speed of light ($c \approx 3 \times 10^5$ km s^{-1}), yields the star's velocity. In this example, the Doppler shift is caused by the star changing its velocity by 12 km s^{-1}.

quantum states of an atom or molecule, the photon can be absorbed. Thus, there are well-defined central wavelengths for every absorption line.[4] If the star moves with respect to us, we'll see the wavelengths of the absorption lines change due to the Doppler effect.

A star's absorption lines can be observed using a spectrometer, which measures the amount of light (flux) from a star in small wavelength intervals. For a typical

---

[4]Absorption lines have widths because of various line-broadening mechanisms. The primary broadening sources in most stars are due to stellar rotation and the motion of hot and cold cells of gas in stellar atmospheres.

spectrometer used to detect planets, light is measured in intervals of about 0.6 nanometer (nm). This might seem small, but consider the wavelength shift of a typical absorption line caused by a Jupiter-sized planet in a one-year orbit. The Doppler shift, z, is given by

$$(2.20) \qquad z = \frac{\Delta\lambda}{\lambda_0} = \frac{v_\star}{c}$$

where $c = 3 \times 10^8$ m s^{-1} is the speed of light, λ_0 is the wavelength where the stellar absorption line is normally centered, and it is shifted by an amount $\Delta\lambda$. Here, I have assumed that the star's motion is negligible compared to the speed of light ($v_\star \ll c$).

For a star moving at $v_\star = 30$ m s^{-1}, a line normally at $\lambda = 600$ nm will be shifted by $\Delta\lambda = 6 \times 10^{-5}$ nm. This is about 10^5 or 10,000 times smaller than the wavelength spanned by a single detector element of a typical spectrometer. Fortunately, every one of the thousands of lines in a star's spectrum encodes the exact same Doppler shift. By measuring the average shift of all of these lines, astronomers can attain the precision necessary to detect planets as small as several Earth-masses in periods of a few days using existing instruments. Future instruments will have even better sensitivity.

However, attaining this level of precision is not trivial. One must not only measure a shift $\Delta\lambda$, but the shift must be measured with respect to a reference wavelength λ_0. Thus, the measurement of highly precise radial velocities relies on a precise mapping of wavelength to pixel within the spectrometer. The spectrometer is a complicated optical instrument. However, it is essentially a prism to disperse light, and at the end of the light path is a simple charge-coupled device (CCD), on which the star's spectrum is recorded for later analysis. The CCD is similar to the detector in a commercial digital camera, and it comprises a grid of small (~15 micron) pixels. Different wavelengths of light are directed onto different pixels. A closely spaced pair of absorption lines can be distinguished as individual lines only if they are separated by more than this resolution element.

The primary problem is the spectrometer is a moving platform. Changes in the ambient temperature surrounding the spectrometer can change the position of optics by a fraction of a wavelength. Changes in pressure can change the index of refraction of the air near the optics. And vibrations can shift the detector with respect to the light path. All of these effects change the mapping of wavelength to pixel within the spectrometer. These changes in the wavelength mapping can easily be mistaken for a Doppler shift of several kilometers per second, which is plenty to mask the much smaller shift due to a planet.

One way to mitigate these false Doppler shifts is to impose a locally generated absorption spectrum onto the stellar spectrum. This can be accomplished by placing a clear, Pyrex glass cell containing a molecular gas at the entrance of the spectrometer. Since the gas cell is not moving, the narrow molecular absorption features serve as a stationary grid, mapping wavelength to pixel, against which the motion of stellar absorption lines can be measured to high precision. While the spectrometer optical components might shift throughout the night, the locations of the lines produced by the gas cell remain fixed in wavelength. This is the technique used at the Keck Observatory with the HIRES spectrometer.

The other method of measuring highly precise radial velocities is to ensure that the instrument does not move. The absorption-cell technique is most often used as a post hoc solution to intrinsic instabilities of general-use spectrometers. However, if a spectrometer is built specifically for precise radial velocities, it can be designed with stability in mind. This stable-platform approach is used on the High-Accuracy Radial-velocity Planet Search (HARPS) spectrometer. HARPS is housed below its telescope and it is enclosed in a temperature- and pressure-controlled environment. The wavelength scale is periodically checked with a lamp that emits light at known, discrete wavelengths. However, because of the intrinsic stability of HARPS, the wavelength scale remains static over the course of a night. For stars requiring the highest possible precision, a separate fiber-optic cable can shine calibration light at the same time that the stellar spectrum is recorded.

2.5 Stellar Jitter

Even with a perfect instrument, the detection of low-mass planets will be complicated by the nonstable nature of stars. While stars can be thought of as spherical blackbodies, in reality they are large balls of gas with no true surface. The "surface" is known as the photosphere, which marks the radius within the star where the gas transitions from opaque to transluscent to outgoing photons. This pseudo-surface, known as the photosphere, can be jostled about by the bubbling, convective gas just below it. The rising convective cells then cause the star to ring at natural vibrational modes that show up as periodic variations in the velocity of the stellar surface. However, since the surface of the star cannot be resolved, all of this jostling results in radial velocity noise that is astrophysical in nature, rather than originating in the instrument. This astrophysical noise is known as "stellar jitter."

In addition to noise sources in the interior of the star, the surface of the star is not uniformly bright. Instead, the surface of stars like the Sun are dotted by dark star spots, bright regions called plage, and the occasional flare. Since stars are approximately spheres, they appear to an observer as a circular disk projected on the sky, with half of the surface approaching the observer and half of the surface receding, due to stellar rotation. The net effect of rotation on an unblemished stellar surface would be for the two halves of the star, one redshifted and one blueshifted, to balance out and no net apparent velocity would be produced. Instead, the star's absorption lines appear to be broadened by the redshifted and blueshifted portions of the stellar surface.

Blemishes on the stellar surface will tip this balance and result in a nonzero net velocity across the stellar surface and absorption lines. For example, a spot on the blueshifted hemisphere of the star would result in a net redshift. Then, as the spot moves across the star's surface following the star's rotation, the net shift would change to a blueshift. This change in the star's surface velocity can be mistaken for the Doppler signal of a planet, with an orbital period equal to the star's rotation period.

The first method of dealing with jitter is to just treat it as an additional source of noise. Even though the processes that give rise to artificial radial velocities produce signals that are time-correlated, if the signal is sampled sporadically—as usually happens due to the vagaries of telescope scheduling—the jitter will look like random scatter. Various calibrations indicate what level of jitter one can expect for a star of given properties. This expected level of jitter can simply be added in quadrature to the measurement uncertainties of the RV measurements, resulting in inflated error bars for each data point.

A better method of dealing with jitter is to use a physical model to fit the jitter signal along with the planet signal. However, this method can only be used when the jitter "signal" is time-resolved with a dense set of measurements spanning the typical timescale for jitter to manifest itself and evolve. For example, jitter due to star spots modulates the RV signal of the star on a timescale of the star's rotation. For Sun-like stars, this rotation timescale is tens of days (24 days for the Sun). An additional feature of this spot-induced jitter is that it is ephemeral, as opposed to the constant signal from the planet. This is because spots come and go on monthly timescales. Thus, if one can sample the star's RV once per night, every night over months, one can in principle see the spot modulation oscillating and fading in and out, along with the underlying, permanent planet signal. For more on this method, I refer the reader to the work of Xavier Dumusque and collaborators (2012), who used high-cadence monitoring of α Cen B to detect a planet (candidate) with a minimum mass ($M_p \sin i$) close to the Earth's mass and that has a Doppler signal below the jitter level of the star.

2.6 Design Considerations for a Doppler Survey

In the previous sections I provided the basic physical picture of a planet tugging on its star, thereby producing a Doppler shift in the star's spectrum. I also demonstrated that the expected signal is quite small, yet measurable with modern spectrometers. In this section I provide examples of how one would design a survey to detect two types of planets. The first case is close-in, gas-giant planets known as "hot Jupiters." These are the easiest planets to detect using the Doppler technique. However, hot Jupiters are found around only $\sim$1% of stars. Thus, the survey approach must maximize the number of stars searched in order to have a reasonable expectation of finding planets.

The second case focuses on Earth-mass planets in the habitable zones of their stars. Compared to hot Jupiters, Earth-sized planets are common throughout the Galaxy. However, because the Doppler amplitude scales proportionally to the planet's mass and inversely with the semimajor axis ($K \propto a^{-1/2} m_p \sin i$), the signals from these planets will be orders of magnitude smaller than the Doppler variations induced by hot Jupiters. Thus, such a survey must be optimized in a very different manner than one searching for more massive planets.

2.6.1 Example 1: Searching for Hot Jupiters

According to Equation 2.10, a Jupiter-mass planet in a 1-year orbit around a Sun-like star induces a signal with an amplitude $K \approx 30$ m s^{-1}. Hot Jupiters orbit much closer to their stars with periods of about 3 days, instead of 365 days. The amplitude scales as $P^{-1/3}$, so a hot Jupiter will induce a signal that is $(P_{\mathrm{HJ}}/P_{\mathrm{AU}})^{-1/3} = (3/365)^{-1/3} \approx 5$ times larger than a Jupiter at 1 AU, or $K \approx 150$ m s^{-1}.

Since modern spectrometers can routinely measure stellar radial velocities with a precision of 3 m s^{-1}, a middle–aged (>1 Gyr), Sun-like star with a hot Jupiter is readily distinguished from a star without one. Even with three to four RV measurements, a star with a hot Jupiter will exhibit RVs with a much larger scatter than it would if it were solitary. This means that only a few RV measurements are necessary to identify stars that are likely to harbor hot Jupiters.

This "quick-look" strategy was first implemented by Debra Fischer with her *Next 2000 Stars* (N2K) survey (Fischer et al., 2005). She and her team surveyed a large number of stars by scheduling telescope time in several three-night observing runs and observing their target stars once per night. For each star's sequence of three RVs, the N2K team calculated the standard deviation of the points, σ_V (top panel of Figure 2.5). Stars that have σ_{RV} larger than some threshold were then followed up intensively to characterize the orbit of the planet candidate. This last step typically required 10–15 additional RV measurements. The N2K survey resulted in the discovery of 17 Jupiter-mass exoplanets, of which two were subsequently found to transit (eclipse) their host stars, demonstrating the efficiency of the quick-look survey strategy. The lower panel of Figure 2.5 shows the RV time series for a planet I helped detect and announce when I was a member of the N2K team as a graduate student.

2.6.2 Example 2: Searching for Habitable-Zone Planets

The habitable zone (HZ) of a star is the range of semi-major axes at which a planet's equilibrium temperature is amenable to the presence of liquid water on its surface. The details of habitability are complicated and varied. Indeed, the topic deserves its own book, and I encourage the reader to see Jim Kasting's *How to Find a Habitable Planet* (2010).

For our purposes, we can think of the HZ as a semimajor axis a_{HZ} characterized by constant planetary equilibrium temperature, T_{eq}, around stars of various masses, $M_\star$. The equilibrium temperature of a planet is reached when the power it receives from the central star is equal to the power the planet emits as thermal radiation. The power per area absorbed at the location of the planet, also known as the flux F with units of energy per time per area, is

$$(2.21) \qquad F_{in} = \frac{L_\star}{4\pi a^2}$$

where $L_\star$ is the luminosity of the star, or total power output. The planet absorbs energy on the day-side hemisphere, which has a projected area πR_P^2 as seen from

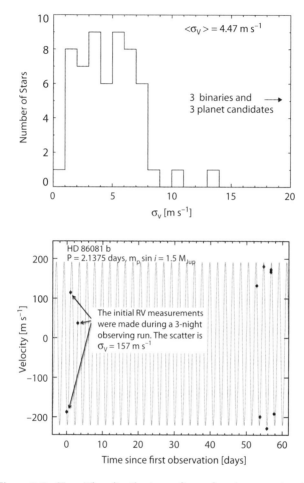

Figure 2.5. *Top:* The distribution of σ_V for the stars in the Keck N2K survey. σ_V is the standard deviation of the first 3–4 RV measurements of each target star. The mean level of RV scatter is only 4.47 m s^{-1}, indicating that most stars in the survey lack a Jupiter-mass planet. A few stars in the sample show very large σ_V, and these stars are subsequently followed up with 10–15 additional RV measurements to confirm and characterize the

the star. Thus, the energy absorbed per unit time by the planet is Area $\times$ (Flux in) $= 4L_\star(R_P/a_{HZ})^2$. Here, I have ignored the amount of light reflected at the planet's surface.

The planet is approximately a blackbody of temperature T_{eq}, and thus also radiates flux over its entire surface, which has a spherical area $4\pi R_P^2$. The flux at the surface of a blackbody is $F_{out} = \sigma T^4$, where σ is the Steffan-Boltzmann constant. The power output is Surface area $\times$ (Flux out) $= 4\pi R_P^2 \sigma T_{eq}^4$. Equating the input and output power and solving for the a_{HZ} gives

$$(2.22) \qquad a_{HZ} = \left(\frac{L_\star}{\pi \sigma T_{eq}^4}\right)^{1/2}$$

Since the HZ is characterized by a constant T_{eq}, the location a_{HZ} is dependent only on the stellar luminosity $L_\star$. For Sun-like stars a good approximation for the relationship between luminosity and mass is given by $L_\star \sim M_\star^4$, which leads to

$$(2.23) \qquad a_{HZ}(M_\star) \sim M_\star^2$$

The Earth resides in the Sun's habitable zone at $a_{HZ}(M_\star = M_\odot) = 1$ AU. A star with 70% the mass of the Sun has $a_{HZ} = 0.7^2 \approx 0.5$ AU.

Figure 2.5. (*Continued*)
planetary orbit. *Bottom:* An example hot Jupiter discovered by the N2K Consortium. The first three points have $\sigma_V = 157$ m s^{-1}, indicative of a planet candidate. The planet was later confirmed and published as HD 86081 b (Johnson et al., 2006).

We can now derive the velocity amplitude of a HZ planet K_{HZ} using Equation 2.11, which results in

$$(2.24) \qquad K_{\mathrm{HZ}}(M_\star) \sim M_\star^{-3/2} m_p \sin i$$

This result indicates that for a given planet mass in the habitable zone, the largest Doppler signals are expected for stars with lower masses since the amplitude scales inversely with $M_\star$. For example, an Earth-sized planet orbiting at a_{HZ} around a red dwarf with $M_\star = 0.3 \, \mathrm{M_\odot}$ will have a Doppler amplitude that is six times larger than the same planet in the HZ of a Sun-like star, or $K_{\mathrm{HZ}} \approx 55 \, \mathrm{cm \, s^{-1}}$, compared to $9 \, \mathrm{cm \, s^{-1}}$ for the Sun orbited by the Earth. Of course, our calculation of the location of the habitable zone includes several implicit, simplifying assumptions. A fuller treatment would consider the effect of the star's spectral energy distribution on the planet's temperature (e.g., Kasting et al., 1993; Shields et al., 2013). However, this simple derivation demonstrates that stellar mass is clearly an important consideration in planning a search for HZ planets.

2.7 Concluding Remarks

In this chapter we examined how planets belie their presence by gravitationally tugging on their host stars, which causes stars to move back and forth along our line of sight. Astronomers can detect this motion by monitoring the velocities of stars using spectrometers to track the Doppler shift of stellar absorption lines. Because the amplitude of the RV variability scales as $K \propto a^{-1/2}$, and because planets have periods less than the observing baseline, the Doppler technique is most sensitive to close-in planets. Presently, the longest-running Doppler surveys have time baselines of about 20 years, and the planet with the longest, precisely measured orbital period is 55 Cnc d, which has $P = 13.44 \pm 0.08$ years.

The majority of planets detected with the Doppler technique have periods less than a year. In fact, one of the shortest-period planets also orbits the star 55 Cnc. The planet, 55 Cnc e, has a period of only 17.76 *hours*! The planet's mass is $m_p = 8.3$ M$_\oplus$, which is unambiguously measured because the inclination is known very precisely: $i = 83.4^{+1.5}_{-1.7}$ degrees. The inclination is known because the planet passes in front of the central star as viewed from Earth. The details of these planet transits are the subject of the next chapter.

3

SEEING THE SHADOWS OF PLANETS

> It should be possible, without much difficulty, to discover planets with the mass 10 times the mass of Jupiter by the Doppler effect. There would, of course, also be eclipses.
>
> — *Otto Struve, 1952*

"Oh no! What happened? Where's the star?!" It was a summer night in 2008 and I was at the controls of the University of Hawaii's 88-inch (2.2-meter) telescope. Well, to be exact, I was at sea level in the beautiful Manoa valley on the island of Oahu remotely operating the telescope, which is two islands over on the Big Island, atop Mauna Kea near the Keck telescopes. I was talking to the telescope operator via a telecon link, and I was very frustrated. When the dust settled (literally), the operator discovered that the telescope dome had automatically moved to a safe position due to excessively high winds. This move protects the delicate primary mirror from abrasion from wind-blown dust; the wind in question was gusting up to 50 miles per hour at the time.

At a logical level, I understood the importance of protecting the telescope. But as a young observational astronomer trying to make a name for myself, I was operating at a very emotional level. All that I knew at the time was that my target star, WASP-10, was no longer visible. The photons that began their journey to the Earth over 300 years ago were bouncing off the telescope dome rather than being counted by the CCD detector. Another telescope night lost; back to the drawing board.

When I arrived at the Institute for Astronomy at the University of Hawaii as a newly-minted postdoctoral fellow, I began casting about for a new research direction. As a student, my job was to serve my thesis adviser and work on my dissertation topic. As a postdoc, my job was to branch out into new scientific directions and establish myself as an independent researcher. This was both an exhilarating opportunity and a considerable challenge. Independence to do what I wanted was nice, to be sure. But I was also left missing the guidance of my adviser. What if I struck off in the wrong direction? What if I overlooked something obvious along the way? A few months into my new job I decided that nothing ventured would result in nothing gained, and being wrong would be no more a net negative than staying put.

After exploring a few directions closely related to my thesis work that mostly involved the radial velocity technique, I decided to try my hand at studying planets that eclipse (transit) their stars as viewed from Earth. In the previous chapter we examined how a star's spectrum changes in response to the gravitational tug of an orbiting planet. In that case, the change to the star's light is observed in the star's spectrum as a Doppler shift in the absorption lines. For orbits that are by chance viewed edge-on, the planet will pass between the observer and its star, thereby blocking a fraction of the star's light. By making repeated measurements of a star's brightness, astronomers can construct a transit light curve. The characteristic shape of a transit light curve encodes information about the physical characteristics of the planet, as well as information about the orbit and even the stellar properties.

The ability to measure the physical and orbital properties of a planet from a transit event depends critically on the photometric precision of the light curve. The photometric precision is based on the ability of an instrument (photometer) to record the same brightness level time after time, such that any deviation from a constant brightness can be trusted as having an actual astrophysical origin rather than resulting from an instrumental effect or the Earth's atmosphere. I had tried my hand at transit light curve observations while I was a student at Berkeley, using the 1-meter Nickel telescope atop Mt. Hamilton at the Lick Observatory in Northern California. While the precision I was able to attain was good by the standards of the time (about 1 part in 1000, or approximately 1 millimagnitude [mmag]; Johnson et al., 2008), I was left with the question: "Why can't anyone do better than this level of precision?"

This particular story ends with my discovery that there is, indeed, a way to do much better than had been done before. My path involved figuring out the physics behind a transit light curve, which in turn motivated the need to attain better precision and helped me identify the important ingredients that would help me get there. The transit light curve that resulted from my pursuit of higher precision is shown in Figure 3.3, demonstrating a precision of better than 0.5 mmag—a record-setter at the time. We now routinely see light curves like this, or orders of magnitude better, for planets discovered by the NASA *Kepler* space telescope. This exceptionally high precision, afforded by observing from space above the deleterious effects of the Earth's atmosphere, has opened up new discovery space and paved the way toward finding planets like our own around other stars.

In what follows we explore the details of the transit light curve and how transits can be used to both discover and characterize planetary systems. In the process, we retrace the steps that led me to measure an exquisite transit light curve of the WASP-10 planetary system, and how the NASA *Kepler* Mission revolutionized the field of exoplanetary science by discovering thousands of new, transiting planetary systems, many of which contain planets resembling our own.

3.1 Measuring and Reading Transit Signals

In most of this chapter, unless otherwise noted, I focus on planets on circular orbits. Just as in the previous chapter, studying circular orbits allows us to trade generality for physical intuition. Further, we will assume that during the eclipse of its host star, the planet moves at a constant speed, and that the planet's radius is small compared to the radius of the star.

It is important to realize that the transit method of finding planets does not detect all planets because not all planets have the necessary geometry to eclipse. At a given semimajor axis a, only a narrow range of inclinations

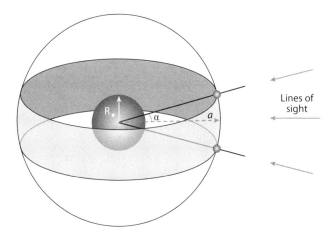

Figure 3.1. Illustration of the "strip" of solid angles in which the planet can lie and transit the star as viewed from the Earth (observer viewing from the far right at a distance $\gg a$). The height of the strip is $\sin^{-1}[(R_\star + R_P)/a]$. The solid angle of the strip is calculated by integrating the height 2π around the star. The probability is the solid angle of the strip divided by the 4π steradians covering the sphere of radius a surrounding the star.

carry the planet in front of a star with a radius $R_\star$. The probability of a planet transiting its star as viewed from the Earth is given by the ratio of the "strip" of angles shown in Figure 3.1 divided by total solid angle over the surface of a sphere: $\Omega = 4\pi$ steradians, or about 4.1×10^4 square degrees.

The area of the strip is straightforward to calculate in spherical coordinates, with the angle α measured from the horizontal,[1] and spanning the angles that intercept the

---

[1]As opposed to the standard θ measured from the vertical in spherical coordinates, as is often the case in physics textbooks.

stellar surface. These angles range from $\alpha_{min} = -\sin^{-1}[(R_\star + R_P)/a]$ to $\alpha_{max} = \sin^{-1}[(R_\star + R_P)/a]$, and $\theta_{min} = 0$ to $\theta_{max} = 2\pi$. The probability is therefore

$$(3.1) \qquad P_{transit} = \frac{\int_{\alpha_{min}}^{\alpha_{max}} \int_0^{2\pi} \cos(\alpha) d\alpha d\theta}{4\pi}$$

where R_P is the radius of the planet, which is included to account for grazing transits when only a portion of the planet is blocking the star. Solving the integral results in

$$(3.2) \qquad\qquad P_{transit} = \frac{R_\star + R_P}{a}$$

As viewed from outside of the Solar System, the Earth has a transit probability of $R_\odot/(1 \text{ AU}) = 7 \times 10^{10}$ cm/1.5×10^{13} cm $\approx 0.5\%$.

This small transit probability (a 1 in 200 chance) highlights one key aspect of a transit survey: the need to survey a large number of stars in order to have a decent chance of detecting a transit. This also demonstrates why hot Jupiters—giant planets with periods less than 10 days—have played such an important role in our understanding of planets in general. Because they orbit roughly 100 times closer to their stars than does Jupiter in the Solar System (smaller a), there are many examples of transiting short-period Jupiters. The typical hot Jupiter has an orbital period of about 3 days. We know from Kepler's Third Law that $a \propto P^{2/3}$, which means that a typical hot Jupiter has a transit probability that is roughly $(365/3)^{2/3} \approx 25$ times higher than that of the Earth around the Sun, resulting in $P_{transit} \approx 10\%$ for a hot Jupiter.

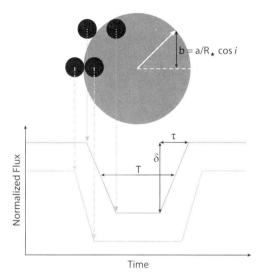

Figure 3.2. Illustration of the transit of a planet (black circles) in front of a star (large gray circle). Two different impact parameters are shown: for $b = 0$ the planet transits along the stellar equator, while for $b > 0$ the planet traverses a shorter chord along the stellar surface, resulting in a shorter transit duration (smaller T) and a shallower ingress/egress slope (longer τ). The depth δ is given by the ratio of areas between the planet and star, or $\delta = (R_P / R_\star)^2$.

Like the Sun and the Moon as viewed from afar, the sky-projected areas of the spherical planet and star appear to be circular disks when projected on the sky. Immediately following the first contact between the stellar and planetary disks, the planet gradually blocks more and more of the star, causing a steady decrease in the star's light. This interval is known as *ingress*, and there is an inverse effect during *egress*, as the planet moves away from the stellar disk. This configuration is shown in Figure 3.2.

To gain a physical intuition for transit light curves, we will make several simplifying assumptions about the star, the planet and their mutual orbit. As stated previously, we'll assume the orbit is circular and that the planet moves across the face of the star at a constant speed during the transit. Further, we assume the planet is a perfectly circular, opaque disk seen against a uniformly illuminated stellar surface. These assumptions neglect the oblateness in the planet, its partially transmissive atmosphere, elliptical orbits, dark star spots, and the characteristic limb darkening that causes stars to appear brighter at their disk centers and fainter toward the edges. These factors all have measurable effects on observed light curves. But ignoring them does not affect the features of a transit light curve that are most important for detecting planets and measuring their properties, at least to first order.

The most widely used analytic equations that describe the time variations observed in a transit light curve were developed by a Caltech undergraduate researcher, Kaisey Mandel, who was working with Eric Agol, a postdoctoral researcher at the time (Mandel & Agol, 2002). Around the same time, Sara Seager, a postdoc at the Institute for Advanced Study in Princeton, and Gabriela Mallén-Ornelas were attacking the same problem (Seager & Mallén-Ornelas, 2003). The expressions in the Mandel-Agol and Seager-Mallén-Ornelas light curve models are fairly complex, but they are based principally upon the geometry of overlapping circles. Rather than examining the transit light curve model in detail, I will instead focus on the observed features and how they relate to the key physical properties of the system. The goal is for the student to be able to read the physics of a transiting planetary system right off a transit light curve.

The schematic light curve shown in Figure 3.2 illustrates the three key observables using the notation from Carter et al. (2008): the ingress/egress duration (τ), the "total" duration (T) and the transit depth (δ). The depth is the amount of starlight blocked by the planet, which is equal to the ratio of areas, or $\delta = (R_p / R_\star)^2$. This relationship may be simple, but it provides a lot of information about the transit technique. First, note that one cannot measure the planet's radius directly from the transit light curve. The observer must know the radius of the star in order to measure the radius of the planet.

The second important aspect of the transit depth equation is that there are two ways to increase the transit signal: bigger planets (larger R_P) or smaller stars (smaller $R_\star$). For example, Jupiter has a radius that is 9.75 times smaller than the Sun's, and its transit would cause the Sun to appear fainter by $(1/9.75)^2 \approx 0.01$, or 1%. The Earth's radius is 109 times smaller than the Sun's, resulting in a paltry transit depth of 8×10^{-5}, or 80 parts per million. At present, only the NASA *Kepler* space telescope can discern such a shallow transit. However, an Earth-sized planet around the nearby, diminutive red dwarf Proxima Centauri ($R_\star = 0.14\ R_\odot$) would cause a transit depth, $\delta = 0.004 = 0.4\%$, which is comparable to that of a Jupiter-size planet around a Sun-like star. This is one reason why astronomers see red dwarfs as particularly good targets for the search for Earth-like planets.

The ingress/egress duration, τ, is related to the speed and size of the planet, and the perpendicular distance between the planet's path and the center of the star, denoted by the impact parameter b (see Figure 3.2). For the special case of the planet transiting along the stellar equator ($b = 0$), the ingress/egress time is

$$(3.3) \qquad \tau = \frac{2R_p}{v_p} = \frac{P}{\pi} \frac{R_P}{a}$$

where v_p is the speed of the planet as it crosses the stellar disk, or $2\pi a / P$. Similarly, the transit duration is just the time needed for a planet moving at a Keplerian speed to traverse the diameter of the star, or

$$(3.4) \qquad T = \frac{2R_\star}{v_p} = \frac{P}{\pi} \frac{R_\star}{a}$$

More generally, the planet crosses the star along a chord that lies along a path offset from the stellar equator, given by the impact parameter, $b = (a / R_\star) \cos i$, where i is the inclination of the planet's orbit, measured from the vertical, where $i = 90$ degrees corresponds to an equatorial transit. For example, the upper light curve in Figure 3.2 is for $b > 0$, $i < 90$ degrees, and the lower light curve is for the special case $b = 0$, $i = 90$ degrees. In this more general case with $b > 0$, the total and ingress/egress times become

$$\tau = \frac{P R_p}{\pi a} \frac{1}{\sqrt{1 - b^2}}$$

$$(3.5) \qquad T = \frac{P R_\star}{\pi a} \sqrt{1 - b^2}$$

Thus, compared to an equatorial transit, those with nonzero impact parameters ($b > 0$) have longer ingress/egress durations and shorter overall durations. In other words, the higher the impact parameter, the more V-shaped the transit light curve appears.

We can now relate the transit observables (δ, τ and T) to the physical properties of the planetary system:

$$\frac{R_p}{R_\star} = \sqrt{\delta}$$

$$b^2 = 1 - \delta^{1/2}\frac{T}{\tau}$$

(3.6) $$\frac{a}{R_\star} = \frac{P\delta^{1/4}}{2\pi}\left(\frac{4}{T\tau}\right)^{1/2}$$

In these expressions, $R_P/R_\star$ is commonly referred to as the planet–star radius ratio and and $a/R_\star$ is the *scaled semimajor axis*.

We are now in a position to read off the transit parameters from the light curve of WASP-10 shown in Figure 3.3. From measurements of successive transits, the period of WASP-10b has been previously measured to be approximately 3 days.[2] The light curve has a round shape at minimum light because of limb darkening: the stellar disk is not uniformly bright, but instead brighter in the middle and fainter at the edge (limb). However, we can approximate the transit depth by measuring the average depth between the end of ingress and the beginning of egress, which results in $\delta \approx 0.027 = 2.7\%$.

[2]The true period is measured very precisely: $P = 3.0926813 \pm 0.000012$ days.

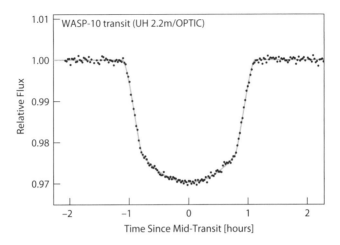

Figure 3.3. A transit light curve measured for the WASP-10 hot Jupiter planetary system (Johnson et al., 2009). Observations were made with the University of Hawaii 2.2-meter telescope with the OPTIC camera.

The total duration of the transit is approximately the full width at half of the maximum depth, which is $T \approx 1.9$ hours (0.079 day). The ingress/egress duration is about $\tau \approx 0.33$ hour (0.014 day). These measurements correspond to the transit parameters

$$\frac{R_p}{R_\star} \approx 0.16$$

$$b \approx 0.3$$

(3.7)
$$\frac{a}{R_\star} \approx 12$$

These parameters are very close to those reported in Johnson et al. (2009), who found $R_p/R_\star = 0.1582^{+0.0007}_{-0.0018}$, $b = 0.299^{+0.035}_{-0.054}$ and $a/R_\star = 11.58 \pm 0.13$.

3.2 The Importance of $a/R_\star$

Note the repeated appearance of the scaled semimajor axis, $a/R_\star$. It shows up in the transit probability, again in the equation for the transit duration, and given the depth, period and durations of the transit it represents a key physical observable of the transiting planetary system. It turns out that $a/R_\star$ is also directly related to the stellar density, as follows.

By using Newton's version of Kepler's Third Law, the semimajor axis can be expressed in terms of the orbital period and stellar mass

$$(3.8) \quad a/R_\star = \left(\frac{G P^2}{4\pi^2} \right)^{1/3} \left(\frac{M_\star^{1/3}}{R_\star} \right) \sim \rho_\star^{1/3}$$

The last term on the right-hand side is the cube root of the average stellar density, $\rho_\star$, and I have dropped the period dependence since the periods of transiting planets are usually measured to a precision of 1 second or better, making P essentially a constant in normal practice. Solving for the density,

$$(3.9) \quad\quad\quad \rho_\star \sim (a/R_\star)^3$$

It is important to note that in this case we have retained our assumption of a circular orbit. This method of estimating the stellar density will be incorrect for a planet on an eccentric orbit. Fortunately, most close-in giant planets are close enough to their stars to have their orbits circularized though tidal interactions with the star.

The ability to measure the stellar density from the transit light curve is an especially important result because it provides a means of estimating the stellar radius, which in turn enables an estimate of the planet's radius from the transit depth. Measuring the radius of a single star is usually done using the well-known Stefan-Boltzmann equation

$$(3.10) \qquad L_\star = 4\pi R_\star^2 \sigma T^4$$

Solving Equation 3.10 for the stellar radius gives

$$(3.11) \qquad R_\star = \left(\frac{L_\star}{4\pi \sigma} \right)^{1/2} T^{-2}$$

The effective temperature of the star, T, can be measured by modeling the observed stellar spectrum. However, in order to estimate $L_\star$, the distance to the star must be measured precisely since photometry gives only the *apparent* brightness of the star. Trigonometric parallaxes can be measured for nearby stars, but the target stars of wide-field transit surveys typically lie too far from the Sun to be measured directly in this manner.[3]

---

[3]This is true for now, with parallaxes available only for bright stars from the *Hipparcos* mission. However, the *Gaia* mission will gather parallaxes for about *one billion* stars. The *Gaia* space telescope is set to be launched in 2013 and will complete its mission in 2018.

Fortunately, the stellar density as measured from the transit light curve can be used as a proxy for the luminosity when combined with stellar evolution models. Normally, stellar evolution models provide an estimate of $R_\star$ when three values are interpolated onto the model grids: T, $L_\star$ and the chemical composition, Z. However, these models can be recast in terms of $\rho_\star$ rather than $L_\star$. In fact, the models are capable of providing a better (less degenerate) solution when expressed in this manner. Thus, transits reveal the physical characteristics of the planet and the star.

For noncircular orbits ($e > 0$) there is an additional factor that depends on e and ω, commonly referred to as $g(e, \omega)$, such that

$$(3.12) \qquad \rho_\star = g(e, \omega)^3 \rho_{\text{circ}}$$

where ρ_{circ} is the density that would be inferred by assuming $e = 0$ and is given by

$$(3.13) \qquad \rho_{\text{circ}} = \frac{3P}{G\pi^2} \left(\frac{\delta^{1/4}}{\sqrt{T\tau}} \right)^3$$

and

$$(3.14) \qquad g(e, \omega) = \frac{1 + e \sin \omega}{\sqrt{1 - e^2}}$$

In cases where the assumption of a circular orbit is not safe (e.g., for planets orbiting far enough from their stars that the tidal circularization timescale is longer than the star's age), one must obtain radial velocities of the star to measure the orbital eccentricity. For brighter transit host-stars radial velocities are routinely obtained in order to measure the mass of the planet, and a self-consistent solution for both the radial velocity orbit and transit light curve can be used to obtain highly precise stellar and planetary properties. Or, if one has a handle on the stellar density, the problem can be inverted to give the planet's eccentricity.

This method had been proposed previously, but had only been used for a statistical assessment of the eccentricity distribution of large samples of exoplanets. However, Harvard graduate student Rebekah Dawson and I showed that one can measure the eccentricity for *individual* transiting systems using this method, which we named the *photoeccentric effect* (Dawson & Johnson, 2012). This is a useful method of measuring the orbital characteristics of transiting planets without investing large telescope time for radial velocity measurements.

3.3 Transit Timing Variations

When searching for planets using the transit method, a single transit event can give away the presence of a planet candidate. However, it is necessary to observe multiple transit events in order to confirm the planet's existence and measure its radius and orbital characteristics. If there exists only a single, isolated planet orbiting the star, then the transits will occur at regular intervals separated by the orbital period. However, the assumption of perfectly periodic orbits doesn't necessarily hold for systems of multiple planets. Two or more planets will feel not only the star's gravitational influence, but there will also be gravitational interactions among the planets in the system.

Consider two transiting planets approaching the point in their orbit where they transit the star (inferior conjunction). If one planet is leading the other, it will feel a gravitational tug from the trailing planet in the direction opposite its orbital motion. This tug will cause the leading planet to arrive to its transit later than predicted. Similarly, the leading planet will pull on the trailing planet, causing it to arrive late to its transit. Thus, for a system of two transiting planets, transit timing variations will be observable for both objects.

However, in many cases in one planet is known to transit its star, and transit timing variations in the known planet can belie the presence of an additional planet or planets in the system. By modeling these variations, the mass and orbital characteristics of the unseen body can be inferred. In this way, transit timing variations can be used as an additional tool for discovering exoplanets.

An example of such a situation is shown in Figure 3.4 for the transiting planet candidate known as *Kepler* Object of Interest number 1474.01, or KOI-1474.01, also known as *Kepler*-419 b. The transits reveal the presence of a planet with a radius approximately equal to that of Jupiter and an orbital period of 69 days. Sequential transits are plotted on the same scale by subtracting $T_{tr} + N$ from the time axis, where T_{tr} is the time of transit for the first event, P is the orbital period and N is the number of the transit event. For a perfectly periodic, single-planet system, each transit event

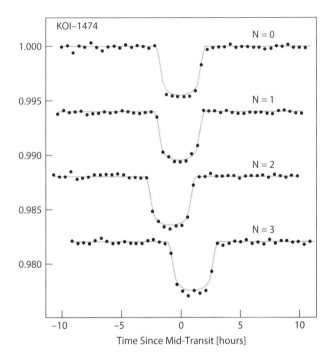

Figure 3.4. Sequential transits of *Kepler*-419 b observed by the *Kepler* space telescope. If the transits were strictly periodic, all of the light curves would be aligned on this scale. However, the Jupiter-sized planet responsible for the transits is experiencing gravitational tugs from an additional planet in the system, which causes the transiting planet to arrive early and late to each transit event.

would line up with the previous. However, in the case of *Kepler*-419, the planet alternates between arriving early and late to the predicted transit time. Modeling these variations provides information about the masses of the planets as well as their mutual inclinations (Dawson et al., 2012).

3.4 Measuring the Brightness of a Star

To this point we have seen how the transit light curve provides information about the planet, its host star and their mutual orbit. However, I haven't yet addressed how astronomers measure these light curves in the first place.

Modern astronomical imaging instruments use a charge-coupled device (CCD) to collect photons from stars. An analogy for a CCD is a large, square parking lot broken into a grid. Each grid cell contains a bucket. If one were to climb a ladder over the lot and burst a giant water balloon, the distribution of water drops would be recorded in the buckets. By measuring the amount of water in the buckets, one would be able to estimate the volume of the water balloon.

A CCD is like a miniature version of this setup, with the buckets replaced by pixels that contain electrons bound to a substrate of silicon. When a photon strikes the silicon pixel, an electron is liberated and trapped in the pixel. At the end of the exposure, the number of electrons are counted in each cell, thereby giving a measure of the amount of light incident on the pixel during the integration time. In this way CCD imagers can be used as photometers that measure the flux (energy per time per unit area) received from stars.

One problem with this approach is that if the CCD is on a ground-based telescope, the Earth's atmosphere is between the instrument and the star. The Earth's atmosphere contains pockets of warm and cool air, which cause streams of photons to be deflected in and out of the line of sight of the telescope. Similarly, patches of the atmosphere may contain small, thin clouds that attenuate the light from stars. For this reason, if a single star's flux is measured with a CCD over the course of a night, the flux level will fluctuate and potentially hide a transit signal.

A way around this problem is to monitor two or more stars simultaneously in the same patch of sky. One star is the target, and the others are the reference stars. Assuming the reference stars do not have planets that transit at the same time as the target star's planetary system, their flux levels should be constant and capable of serving as a photometric reference. In practice, the flux from the target star is measured relative to the flux of the reference stars such that $F_{\text{rel}} = N_T/N_R$, where F_{rel} is the relative flux of the target star, and N_T and N_R are the number of photons measured from the target and reference stars, respectively.

The difficulty is that the host star of a transiting planet is usually the brightest star in a given patch of sky. Normally this is an advantage, because the more photons one gathers, the higher the signal becomes compared to the noise, which is known as the signal-to-noise ratio, or SNR. However, measuring the ratio F_{rel} to high precision depends on having high SNR not only in the target star (N_T), but also in the reference stars (N_R).

As an example, consider the first transiting planet, a hot Jupiter discovered around the Sun-like star HD 209458. The host star has a visual magnitude of approximately $V = 7.6$. While this is beyond the range of brightnesses visible to the naked eye, this star is very bright and therefore very rare on the sky. In fact, there is on average only about one star this bright or brighter per square degree. CCD imagers on most telescopes can capture only about 0.25 square degree at a time, and imagers that can view a full square degree are very rare and typically unavailable for observing transit light curves—an activity that requires most of a precious observing night.

This conundrum can be understood by noting that the uncertainties in the number of photons are governed by counting (Poisson) statistics, such that the uncertainty in measuring N photons is $\sigma_N = \sqrt{N}$. This means that the uncertainty in F_{rel} is given by the propagation of errors in N_T and N_R, or $\sigma_{F_{\text{rel}}}/F_{\text{rel}} = \sqrt{(\sigma_{N_T}/N_T)^2 + (\sigma_{N_R}/N_R)^2} = \sqrt{1/N_T + 1/N_R}$. Thus, the fractional measurement uncertainty in the relative flux is limited not only by the number of photons from the target star, but also by the number of photons counted from the reference stars. If the reference stars provide significantly less flux than the target star, then the measurement uncertainties will be dominated by Poisson fluctuations in the reference stars no matter how much light is collected for the brighter target star.

Naively, one might think that the solution is to just take longer exposures to gather more photons from the target star. However, this approach presents two problems. First, the transit event takes place over a fixed, relatively short amount of time, typically a few hours for a hot Jupiter. So the transit light curve will contain only a few points if one uses, say, hour-long exposures. Also, such long exposures will smear out the transit shape. The second problem with longer exposures is that the target star will eventually saturate the detector. CCD pixels have a finite capacity for electrons, beyond which electrons begin spilling into neighboring pixels. Also, for many detectors the linear relationship between the number of electrons and number of incident photons no longer holds as the pixels approach saturation.

How about taking a large sequence of short exposures? This approach presents three problems. First, at a certain point exposure times become shorter than the time needed to read out the CCD. Modern detectors are becoming faster all the time, but typical CCD read times range from 10 to 90 seconds, depending on the size and age of the detector available at a given telescope. Having a readout time comparable to the exposure time means that half of the transit observing sequence is spent reading out the detector rather than gathering light! Second, CCDs have electrical noise sources that are unimportant at large photon counts, but can become dominant when the number of photons is small, as would be the case for the fainter reference stars when exposure times are decreased. Finally, while the effects of the Earth's atmosphere become less important with longer exposures, these atmospheric noise sources become important for short exposures.

The problem of high-precision photometry with ground-based telescopes comes down to dynamic range (observing both the bright target and fainter reference stars with high SNR) and duty-cycle (spending as much time as possible gathering photons from the star). My solution to these problems was to use a prototype of a new type of detector called an orthogonal transfer array, which was designed and built by John Tonry at the Institute for Astronomy.

Normal CCDs are read out by shuffling the charge from one column to the next. The electrons in the last column are then "dumped" into a readout register, where the electrons are counted by measuring the voltage in each pixel. An orthogonal transfer array can shift charge very efficiently from pixel to pixel during an exposure. For bright stars, this prevents them from saturating, because as a pixel fills up, the charge can be shuffled to neighboring pixels and stored before readout. I programmed the detector to slide the charge around in a square pattern, which allowed me to gather more light in my target and reference stars before I read out the entire chip. This allowed me to collect an order of magnitude more photons for each exposure by using pixels that would otherwise go unused, while using exposure times that were much longer than the readout time. The most successful application of the orthogonal transfer array resulted in the light curve shown in Figure 3.3.

3.5 Radial Velocities First, Transits Second

The observation of a planet transit was made for the HD 209458 system, which was observed in late 1999 by two groups using small, ground-based telescopes (Charbonneau et al., 2000; Henry et al., 2000). The transits of this planet were not discovered as part of a blind survey. Rather, both teams were cleverly increasing their chances of detecting a transiting planet by targeting stars that were already known to harbor hot Jupiters based on radial velocity measurements (Mazeh et al., 2000). If one knows the planet exists, there is an $R_\star/a$ chance that the planet will transit. With enough known planetary systems, the probability that one of them will transit will eventually reach unity.

This method of detecting planets with radial velocities first, and then observing the star for transits second is a tried and true method of finding transiting exoplanets. In fact, the brightest stars known to host transiting planets were all RV targets first. The radial velocity orbit not only shows that a planet exists, but the orbital solution also provides a range of times during each orbital period when the planet might transit its star. This "transit window" is during inferior conjunction, and it coincides with the point in time when the star's radial velocity is zero, and the derivative of the radial velocity is negative. Under ideal conditions, e.g., no measurement uncertainties, the width of the window is simply the transit duration plus about an hour on either side to ensure that the out-of-transit light level is measured accurately. However, because of errors in the period and midtransit time, transit windows can range in width from a factor of a few times the transit duration up to a significant fraction of a day. Despite this uncertainty, spending a night or two searching the transit window of a known exoplanetary system is one of the most efficient means of finding transiting planets because the planet is already known to exist.

In addition to being highly efficient, searching for transits among Doppler-detected planets results in transiting planets around very bright stars. This is because radial velocity surveys are biased toward brighter stars because of the need to disperse the observed starlight into a high-resolution spectrum. On the other hand, wide-field transit surveys are generally biased toward fainter stars because of the requirement of observing many stars to find a single transiting planet. The requirement for more stars translates into the need for a larger search volume, larger distances to the target stars and, as a result, fainter stars. To date, nine Doppler-detected planets were later found to transit, and six of these are among the top-10 brightest transiting systems known.

The brightness of these systems makes them ideal for other types of follow-up observations. However, because the RV-first method is limited to such bright stars, the absolute number of targets is reduced compared to planet searches that detect transits first and use radial velocities for confirmation. This second method of finding planets is the topic of the following section.

3.6 Transit First, Radial Velocities Second

To find a transiting planet, a planet hunter needs to give herself plenty of opportunities to do so. As we saw previously, the probability of a planet transiting its star is given by the geometric relationship $R_\star / a$. For a typical "hot Jupiter" with a three-day orbital period, $a \approx 0.05 \, \text{AU}$. There are approximately 210 solar radii in an astronomical unit, so for a Sun-like star this is $a / R_\star \approx 10$, or a 10% transit probability.

RV surveys have demonstrated that only 1% of stars have a hot Jupiter with a mass greater than a tenth of a Jupiter mass orbiting with a period $P < 10$ days (Wright et al., 2012). Assuming stars with masses and radii similar to the Sun's, the typical transit probability of hot Jupiters is about 10%. Thus, one would need to search $1/(0.01 \times 0.1) = 1000$ stars to expect a single transit!

RV surveys typically target stars with apparent magnitudes down to a V-band magnitude of about $V = 8$. As a point of reference, a star with a magnitude $V = 6$ is at the limit of naked-eye visibility from a dark sight. In comparison, a $V = 8$ star is roughly $2.5^{-2} = 1/6.25$ times as bright as a $V = 6$ star,[4] which is still relatively bright as far as astrophysical objects go. However, the number of stars this bright is only a few thousand over the entire sky, making them very rare and unlikely that multiple $V = 8$ stars will occupy the same field of view.

To increase the number of targets searched by a transit survey, thereby increasing the chances of detecting a large number of planets, surveys must observe fainter stars, which are more numerous on the sky. As a result, planets discovered by transit surveys tend to be much fainter than RV-detected planets, which makes RV follow-up harder. However, RVs are key to confirming the planetary nature of the transit and ruling out false-positive scenarios, and the Keplerian orbit provides the mass of the planet. The mass and radius from the transit depth provide the planet's bulk density.

The expense incurred by the longer exposure times for spectroscopic observations of fainter stars is partially offset by the reduced number of RVs needed to accurately measure an orbit. While RV planets require a minimum of ~10 observations for a secure orbit because of the six

[4]The fluxes of two stars, F_1 and F_2, is related to the stars' respective magnitudes, V_1 and V_2, by $F_1/F_2 = 10^{-0.4(V_2-V_1)} \approx 2.5^{(V_1-V_2)}$.

free parameters (period, eccentricity, time of periastron passage, argument of periastron, Doppler amplitude and velocity offset; see Chapter 2), a transiting planet has a known period and phase, and many hot Jupiters have tidally circularized orbits, thereby requiring only two free parameters (Doppler amplitude and velocity offset) and, at least in principle, the requirement of only only three to four RVs to characterize the system. However, if the signal of the transiting planet is small compared to the measurement precision, or if additional planets are in the system with periods close to that of the transiting planet, more than three to four RVs will be required to securely confirm and characterize the transiting planet. Of course, having more than one planet is a system is a relatively nice problem to have, especially if the multiple planets exhibit TTVs that give additional information about the physical and orbital properties of the system.

To monitor a large number of stars to get a high probability of a transit detection, most transit surveys use telescopes with very wide fields of view. For comparison, while the instrument with the widest field of view on the 10-meter Keck telescope provides about 0.01 square degree, the telescopes used by the Hungarian Automated Telescope Network (HATNet) transit survey cover 68 square degrees in a single pointing. However, each telescope has an aperture of only 0.06 meter, compared to the Keck telescope's 10-meter primary mirror. When designing telescopes, there is typically a trade-off between the size of the aperture and the field of view. But in addition to the wide field of view, another advantage of using small telescopes is their price. The HATNet "telescopes" are often commercial SLR camera lenses, which are extremely inexpensive compared to most astronomical telescopes. At the time of writing this book, ground-based, wide-field transit surveys had produced about 135 new planet detections.

3.6.1 Transit Surveys from Space

Ground-based measurements are hampered by several effects related to the Earth. First, the atmosphere sits between the star and the telescope. Temperature variations in the atmosphere lead to changes in the index of refraction of air, which bends light in and out of the line of sight. This is why stars twinkle. This twinkling is also known as *scintillation*, and it makes measuring the true light level of a star difficult. Another problem related to the Earth is that it is not a stable platform. Its rotation prevents continuous monitoring of the star's brightness due to the day–night cycle and the inability to monitor stars during the day.

Monitoring stars from a space telescope circumvents these problems. The only space telescope dedicated to a transiting exoplanet survey is NASA's *Kepler* survey. Launched in the spring of 2009, the *Kepler* space telescope can be thought of as a 1-meter, wide-field camera. The field of view is 10 by 10 degrees and contains millions of stars, of which about 190,000 were monitored for transits. *Kepler* is beautiful in its simplicity. Once it reached its Earth-trailing orbit, it ejected the dust cover that protected the mirror during launch. At that point, it turned its unblinking eye to a single target field near the constellation Cygnus, just above the galactic plane.

Despite its simplicity, the *Kepler* photometer is capable of measuring the brightness of stars to about 20 parts per million. The combination of this extraordinarily high photometric precision, continuous view and large number of targets means that not much escapes *Kepler*'s notice. To date, *Kepler* has discovered 2740 *Kepler* Objects of Interest, or KOIs. KOIs are stars that show transitlike signals or other interesting transit phenomena, yet await confirmation as actual planets and planet candidates. These transit events remain suspect until ground-based observations can verify that the transit dip is caused by a planet rather than, say, a background-eclipsing binary blended with a brighter foreground star. False positive events plague the ground-based transit surveys, but should only account for less than 10% of the KOIs (Morton & Johnson, 2011; Fressin et al., 2013).

The *Kepler* Mission's many accomplishments and discoveries include the first circumbinary planet (think of the fictional planet Tatooine from the movie *Star Wars*), the first Earth-sized planets, the first sub-Earth-sized planets (one the size of Mercury), and an Earth-sized planet in the habitable zone of its star (Quintana et al., 2014).

Unfortunately, in 2013 the primary *Kepler* Mission came to a sudden end when one of the components that allowed the telescope to point precisely malfunctioned. The telescope was launched with four "reaction wheels" that use the conservation of angular momentum to keep the telescope in a set orientation. Three reaction wheels were used to keep the telescope from rotating around the telescope's three axes, and one wheel was kept in reserve. The first wheel malfunctioned in 2012, and the reserve wheel failed a year later. Fine pointing control is one of the keys to high photometric precision, and the mission was ended since it could not achieve its primary mission objectives.

Fortunately, engineers at Ball Aerospace and NASA came up with a clever solution. Along the back of the *Kepler* spacecraft are mounted two sets of solar panels, which form a ridge. By orienting this ridge into the stream of solar photons, the telescope can be balanced like the bow of a boat pointed into the stream of a river. Since this balancing act is performed at an unstable equilibrium point, the spacecraft's thrusters can be applied at intervals to correct the telescope's fine pointing as it slowly drifts out of equilibrium. This reborn *Kepler* Mission is known as the two-wheel extended mission, or K2 (Howell et al., 2014). The slow drift of the telescope still produces artifacts in the resulting photometric measurements, but these effects can be corrected in software (Vanderburg & Johnson, 2014).

Because the telescope relies on solar radiation to point precisely, the telescope can no longer stare continuously at a single target field. As it orbits the Sun, the telescope must be regularly repositioned to keep the solar panels aligned with the Sun's stream of photons. The benefit of this need to repoint is that every 75 days the telescope can monitor a new field. What it gives up in its previous long time baseline, it gains in the ability to monitor many fields throughout its extended mission. At the time of this book's writing, two new planetary systems have been discovered using K2 photometry (Vanderburg et al., 2014; Crossfield et al., 2015), with the promise of many more discoveries to come as the *Kepler* Mission lives on.

3.7 From Close In to Further Out

The two techniques that I have presented to this point are most effective in detecting planets that are close to their host stars. The Doppler signals of planets scale inversely with semimajor axes, and shorter orbital periods require observations made over shorter durations. The transit probability also increases the closer a planet orbits from its star. Before the first exoplanets were discovered, the expectations for planets around other stars were set by the planets of our Solar System. In this sense, the hot Jupiters and other close-in planets that make up the vast majority of known exoplanetary systems can be thought of as "bonus planets." Without them, our knowledge of exoplanets would be far more limited compared to what we know today.

However, while these planets represent the majority of all known planetary systems, they do not provide a representative picture of planetary systems throughout the Galaxy. To gain a fuller view of the orbital architectures that result from the various planet-formation mechanisms at work throughout the Universe we need methods that are sensitive to planets at much wider separations from their host stars. This leads us to the microlensing and direct imaging techniques in the following chapters.

4

PLANETS BENDING SPACE-TIME

> Thus the results ... can leave little doubt that a deflection of light takes place in the neighbourhood of the sun and that it is of the amount demanded by Einstein's generalised theory of relativity, as attributable by the sun's gravitational field.
> —*F. W. Dyson, A. S. Eddington and C. Davidson, 1920*

Observers using the Keck telescope are occasionally interrupted for special requests known as target-of-opportunity observations (ToO). When a ToO interrupt is made by another observer who is not at the telescope, the astronomer at the telescope must finish the current observation and move the telescope to the requested target. This provision in the allocation of telescope observing time is made because some astrophysical events cannot be predicted at the time that a proposal is written. Supernova explosions and gamma ray bursts happen on their own schedule, not the schedule of the astrophysicists who wish to observe them.

These unpredictable events are known as *transients*, because they happen sporadically and usually occur only once. Another type of transient event is the brightening caused by the coincidental alignment of two stars in the Galaxy as viewed from the Earth, known as *microlensing* for reasons that will become clear later in this chapter.

Microlensing events are fundamentally different from the chance alignments that result in eclipses or transits. While transits result in stars becoming dimmer due to blockages of light, microlensing results in the temporary magnification of starlight. With just the right sort of alignment, the gravitational field of a foreground star (the lens) can bend the path of light from a background star (the source), causing more of the background star's light to pass into the observer's line of sight.

Indeed, the Sun can serve as a lens for background stars, and Einstein's general theory of relativity predicted that the apparent positions of stars near the Sun in the sky could be observed to change during a solar eclipse (overwhelmingly bright sunlight makes observations of nearby stars impossible during other times). A British astronomer named Arthur Eddington led an expedition to the tiny island of Príncipe to observe the total solar eclipse of 1919, primarily to test this key prediction of general relativity. Eddington published his results the following year, providing key observational validation of Einstein's theory.

Einstein was aware that the bending of starlight could, in principle, be observed for other stars in the Galaxy. Rather than simply deflecting the positions of background stars, enough of the light could be bent into the observer's line of sight that the background star would appear brighter. However, Einstein did not immediately pursue the concept. Instead, others developed the theory in the years leading up to 1936, when an engineer named Rudi W. Mandl visited Einstein to talk to him about the microlensing phenomenon (Einstein, 1936). In a letter to

the magazine *Nature*, Einstein wrote "R. W. Mandl paid
me a visit and asked me to publish the results of a little
calculation, which I had made at his request. This note
complies with his wish." In that short note he derived the
geometry and physics of microlensing, but concluded that
there was "no great chance" of ever observing the effect
because it is so unlikely to find two stars in the Galaxy so
close together on the sky (Einstein, 1936).

The chance of a galactic foreground star passing close
enough to a background star is, indeed, extremely small.
The space occupied by a star is tiny compared to the
immense distances between neighboring stars. We can
think of a lens event as a collision that occurs when two
stars are close together on the sky. The cross section for
such a collision, σ, is set by a physical dimension known as
the *Einstein radius*, R_E, such that $\sigma_{\text{lens}} = \pi R_E^2$. The light
from a background star passes within R_E of a foreground
star, then the interaction takes place.

Within this framework, we can assess the order-of-
magnitude probability that a microlensing event will occur
at any specific time by considering the fractional area of
the sky covered by Einstein rings. Imagine a volume of
space within the Galaxy that has an area A projected on
the sky[1] and a depth D_S, which is the distance from the
observer to the available background stars, such that the
volume $\mathcal{V} = D_S A$. The number-density of stars $n_\star$ within
this volume will correspond to a total number of stars on

---

[1]A natural question to ask at this point is, "But do we know the value of A?"
As I tell my students in these situations, fear not! A probability (or rate) will end
up being measured per unit area, so we can trust that A will cancel out as we
move through the derivation. This comes up often in astronomy.

the sky, $N_\star = n_\star D_S A$. Each of these $N_\star$ stars will have an associated cross-sectional area for lensing. The ratio of all of these lensing cross sections to the total area on the sky, A, gives the probability, P_{lens}, of a lensing interaction occurring at any given moment:

$$(4.1) \quad P_{\text{lens}} = \frac{n_\star D_S A \sigma_{\text{lens}}}{A} = n_\star (\pi R_E^2) D_S$$

We can see that this relationship makes sense if we consider the scaling with various parameters. If the density of stars is higher, the probability of a lensing event is correspondingly higher. Similarly, if the cross section for lensing is larger, so too is the lensing probability. Finally, if we look further through the Galaxy we see more stars and end up with a higher lensing probability.

As we will see in the next section, a typical star in the Galaxy has an Einstein radius of about 2 AU, or $R_E \approx 10^{-5}$ pc. To get a rough sense of scale, we can estimate $n_\star$ using the local solar neighborhood. Within 2 pc of the Sun there are five stars: the Sun; the α Centauri triple system; and Barnard's Star. Thus, there are five stars within a volume of 33 pc^3, which corresponds to a stellar density $n_\star = 0.15$ pc^{-3}. Most source stars reside at the distance to the galactic center ($D_S \sim 8000$ pc), resulting in $P_{\text{lens}} \approx 4 \times 10^{-7}$, which means that the chance of a microlensing event is less than one in a million. This is about the same order of magnitude of a more careful calculation based on the observed distribution of stars in the Galaxy (Paczynski, 1991).

Thus, microlensing events are very improbable, and as a result one would need to regularly monitor the brightnesses of hundreds of millions of stars in order to build a sizable sample of events per year. In Einstein's day, such a task was absolutely daunting. Large fields of view could be monitored with telescopes feeding light to photographic plates. However, the analysis of photographic plates was an extremely time-consuming process, as was measuring stellar flux levels of individual stars on each plate. Both tasks had to be done by hand and by eye, which is impossible for millions of stars per exposure observed every clear night.

However, with modern digital detectors attached to robotic, wide-field telescopes, hundreds of millions of stars can be observed each night and their brightnesses measured in an automated fashion by powerful computing clusters. With such a large number of stars observed each night, events with small probabilities can have a decent chance of occurring. In the following section I present the basic geometry of a microlensing event and derive the salient features of a microlensing light curve.

4.1 The Geometry of Microlensing

For the situation in which the source and the lens lie directly along the line of sight, the image of the background star is a ring. The right panel of Figure 4.1 shows an Einstein ring of a source star lensed by a foreground lens star. (Note that in reality, the lens would typically be much fainter than the background source.) The radius of this ring is the Einstein radius, R_E, which has a corresponding angular radius θ_E when viewed from a distance D_L. If the

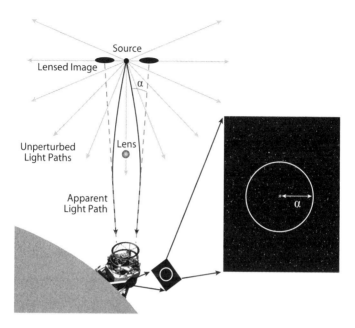

Figure 4.1. Illustration of the paths of light emitted from a source star in the presence of the gravitational field of a lens star. The source emits light isotropically, and only a small fraction of photons travel along paths that end at the observer's telescope. However, some photon paths are bent by the source onto new paths that can be observed. If one traces a straight line back along the end of the photon's path, it does not end at the source. Instead, the paths end at an apparent source position, forming an image. If the source and lens are perfectly aligned, then the image takes the form of a ring, which is shown at right, but with an exaggerated scale. The same lensing effect can be observed for stars lensing light from other stars. However, the angular size of the lensed image, α, will be too small for even the largest present-day telescopes to resolve.

stars are slightly misaligned, there are two distorted images of the background star: a minor image and a major image, with the minor image interior to the Einstein ring radius and the major image exterior to R_E.

During a lensing event, the observer sees the light that would normally be observed from the source star, plus additional light that has been bent into the observer's line of sight. The sum total of the light from the source star and its lensed images is what results in the magnification of starlight. Lensed images produced on angular scales large enough to resolve with telescopes (scales of arcseconds) are known as *macrolensing*, while images produced on scales too small to be observed—typically on the scale of microarcseconds (μas)—are known as *microlensing*.

If a planet orbiting the lens star orbits with a semimajor axis that is comparable to R_E, then the planet may act as a secondary lens. Practically, it is easiest to think about this as the planet lensing the light from the source's image. In other words, planets in just the right configuration can be thought of as lensing the lensed image of the background source star. This secondary brightening—or dimming, depending on the exact configuration—can be used to detect and characterize distant planetary systems.

The physics of microlensing is fundamentally a relativistic effect, requiring an understanding of the details of general relativity. However, lensing can be derived within a classical, Newtonian framework, providing a sense of how various physical parameters result in a microlensing light curve. Consider the path of a photon passing near a lens star of mass M, such that the closest approach between the photon and the star is b, which is known as the *impact parameter*, similar to the variable used to characterize transiting systems. The photon can be imagined to experience a gravitational force directed toward the lens star. The component of this force parallel to the sky plane will induce a change in the photon's velocity with a magnitude $\Delta v \approx (GM_{\text{lens}}/b^2)\Delta t$. Here, I have played it fast and loose by replacing the derivative dv/dt with $\Delta v/\Delta t$ and assuming the force perpendicular to the photon's path is constant during the interaction. However, these assumptions can be shown to be justified.

Now let's consider the time of interaction, Δt. As long as we're using approximations, let's assume the photon feels the gravitational force over a propagation distance $2b$ while traveling at a speed c. With this approximation, $\Delta t = 2b/c$. Since the deflection angle will be small, the force will result in a deflection angle $\alpha = \Delta v/c$, or $\alpha = 2GM/bc^2$. Note that this heuristic derivation is smaller by a factor of 2 compared the result of a more formal derivation involving relativity, which gives

$$(4.2) \qquad\qquad \alpha = \frac{4GM}{bc^2}$$

Missing by a factor of 2 isn't bad given the Newtonian treatment of an intrinsically relativistic effect!

Investigation of Equation 4.2 shows that as the mass of the lens increases, so too does the deflection angle, since space-time is bent more severely by more massive objects. If the source and lens have a smaller impact parameter, b, then the lensing will be stronger. Finally, as $b \to \infty$, $\alpha \to 0$.

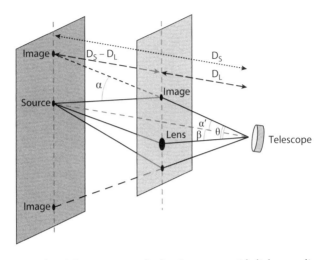

Figure 4.2. The geometry of a lensing event, with light traveling from the source (left) past the lens and ending at the observer's telescope (right). The polygons illustrate the respective sky planes of the source and the lens. The angles between the lens and source (β), the lens and image (θ), and the source and image (α') are labeled accordingly. The distance from the observer to the lens is labeled D_L, and the distance from the observer to the source is D_S. The deflection angle of the photon due to the gravitational influence of the lens is label α.

Now consider a lens star at a distance D_L away from the Earth and a source (background) star at a distance $D_S > D_L$, as shown in Figure 4.2. A typical source star lies at the center of the Galaxy, so $D_S \approx 8000$ pc, while the typical lens is about halfway to the galactic center at $D_L \approx 4000$ pc. If the sight line is toward the source star, the source and lens have an angular separation β. The resulting image—in the figure, the major image is shown—will be an angle θ away from the lens, and an angle α from the source.

The *lens equation* relates these various angles, giving $\beta = \theta - \alpha'$, where α' is distinct from α in Equation 4.2. Inspection of Figure 4.2 shows that α and α' are related through basic geometry:

$$(4.3) \qquad \alpha' = \left(\frac{D_S - D_L}{D_S} \right) \alpha$$

In this expression I have assumed small angles, which results in $\alpha/(D_S - D_L) \approx \alpha/D_S$ because both $D_S - D_L$ and D_S are very large.

Using these angles, along with Equation 4.2, the angular separation between the source and lens can be expressed as

$$(4.4) \quad \beta = \theta - \alpha' = \theta - \frac{4GM}{\theta c^2} \left(\frac{D_S - D_L}{D_S D_L} \right)$$

where I have replaced b with θD_L. Note that when $\beta \to 0$, i.e., when the source and the lens both fall along the line of sight, we can solve for a special value of θ. Note that all light paths around the lens will be bent equally by this amount, thereby forming a ring that has an angular radius

$$\theta_E \equiv \left[\frac{4GM}{c^2} \left(\frac{D_S - D_L}{D_S D_L} \right) \right]^{1/2}$$

$$(4.5) \qquad = \left[\frac{4GM}{c^2} \left(D_L^{-1} - D_S^{-1} \right) \right]^{1/2}$$

Following Gaudi (2011), we can rewrite Equation 4.5 in terms of a relative parallax[2] $\pi_{\rm rel} = {\rm AU}(D_L^{-1} - D_S^{-1})$, and a new term $\kappa = 4G/(c^2{\rm AU}) = 8.14\,{\rm mas}\,M_\odot^{-1}$, which leads to

$$(4.6) \qquad \theta_E \equiv (\kappa\,M\pi_{\rm rel})^{1/2}$$

For a sense of scale, a typical source would be near the galactic center at $D_S = 8$ kpc, where kpc stands for *kiloparsec*, or 1000 parsecs. A typical lens is halfway to the galactic center, or $D_L = 4$ kpc. These typical distances correspond to $\pi_{\rm rel} = 125\,\mu$as. We can now derive a more useful, quantitative relationship for the angular Einstein ring radius in terms of typical values for M, D_L and D_S:

$$(4.7) \quad \theta_E = 1\,{\rm mas}\left(\frac{M}{M_\odot}\right)^{1/2}\left(\frac{\pi_{\rm rel}}{125\,\mu{\rm as}}\right)^{1/2}$$

In deriving the time-variable effect of the lens star on the light from the source star, it is helpful to express the various angles in units of θ_E. This gives two new variables, $u = \beta/\theta_E$ and $y = \theta/\theta_E$. Recall that β is the angle

---

[2]Parallax is the apparent motion of an object on the sky due to the physical motion of an observer. An observer on the Earth moves a distance of 1 AU as the Earth moves in its orbit during a quarter of a year. A star at a distance of 1 pc viewed at the beginning and end of this period of time will appear to move an angular distance of $\pi = {\rm AU/pc} \approx 5 \times 10^{-6}$ radians with respect to very distant background stars, which appear static. There are $206{,}625 \approx 2 \times 10^5$ arcseconds per radian, so the parallax of a star at 1 pc viewed from a baseline of 1 AU is 1 arcsecond, hence the word *parsec*. In this chapter, most distances are on scales of 1 kpc = 1000 pc. A star at a distance of 1 kpc will have a parallax that is 10^3 times smaller than a star at 1 pc, or $\pi = 10^{-3}$ as = 1 mas.

between the source and the lens, so u is this angle in units of θ_E. Similarly, y is the scaled angle between the lens and the image. Substituting these variables into Equation 4.4 gives the simplified relationship

$$u = y - y^{-1}$$

$$(4.8) \qquad y^2 - uy - 1 = 0$$

We're interested in the locations of the major and minor images located at y_+ and y_-, respectively. These angles are the roots of Equation 4.8:

$$y_+ = \frac{1}{2}\left(\sqrt{u^2 + 4} + u\right)$$

$$(4.9) \qquad y_- = -\frac{1}{2}\left(\sqrt{u^2 + 4} - u\right)$$

As an extension of energy conservation, the amount of light per unit solid angle is a conserved quantity. As a result, when the distorted images of the source star contribute additional light along the line of sight, the source appears to increase in brightness, which is known as magnification. Quantitatively, the magnification is given by the ratio of $y_\pm\, dy$ to $u\, du$ (see Figure 4.3), such that

$$(4.10) \qquad A_\pm = \left| \frac{y_\pm}{u} \frac{dy_\pm}{du} \right|$$

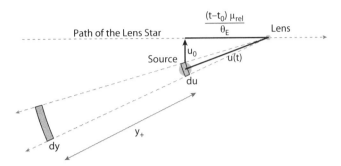

Figure 4.3. Two effects are illustrated in this figure. The solid black lines show the relationship between the path of the lens (left to right along the dashed, horizontal line) and the scaled distance between the source and lens ($u(t)$) and the minimum scaled impact parameter u_0. The magnification scales approximately as $A \propto 1/u(t)$. One of the lensed images of the source is located at y_+, and is magnified because the surface brightness of the source is conserved. The image of the source is compressed in the radial direction by dy_+/du and extended tangentially by y_+/u, resulting in a magnification given by Equation 4.11.

Differentiating Equations 4.9 with respect to u provides $dy_\pm/du$. Substituting Equations 4.8 and 4.9 into Equation 4.10 gives the total magnification as a function of time:

$$(4.11) \quad A(t) = A_+ + A_- = \frac{u(t)^2 + 2}{u(t)\sqrt{u(t)^2 + 4}}$$

This function formally goes to infinity as $u \to 0$, but this occurs only for infinitely small lenses. Actual lenses have a finite extent, so infinite magnification is never reached.

However, we can take the series expansion of $A(u)$ near $u = 0$ to see how the magnification behaves for small values of u. The first-order term of such an expansion shows that $A(u) = 1/u$ for $u \rightarrow 0$, which is a useful tool for an eyeball evaluation of a microlensing light curve.

4.2 The Microlensing Light Curve

To understand the time dependence of the magnification during a lensing event, consider the lens and the source stars as they move through the Galaxy. The stars will not remain static with respect to the line of sight, but will instead have a nonzero relative speed v_{rel}, which will appear as a relative angular motion μ_{rel} on the sky. This sky motion is known as *proper motion* and is typically measured in units of milliarcseconds (10^{-3} arcseconds; denoted by mas) per year. However, we need to be sure to use radians per second when doing calculations.

As a result of this relative proper motion, the projected distance between the objects, u, will change with time. The stars will first move closer, causing u to decrease toward some minimum value, u_0 when the stars are closest together on the sky. At this time the magnification will be at a maximum. This is easy to see for the case that u_0 is small because $A \rightarrow 1/u$ as $u \rightarrow 0$. As the stars pass each other and move further away, A will decrease in a manner symmetric with the rise in magnification.

The time dependence of the projected separation, $u(t)$, can be derived by examining the geometry of the configuration shown in Figure 4.3. In the figure, the source is at the origin and the lens moves from left to right. Recall that all angles are measured as a fraction of the Einstein radius, which is not illustrated in the figure. The objects have a relative speed of μ_{rel}, and let all events be measured at a time t measured with respect to the time of closest approach, t_0. The perpendicular distance traveled by the lens from u_0 is $\mu_{\text{rel}}(t - t_0)$. Geometry gives the separation, u, as a function of time:

$$(4.12) \quad u(t) = \sqrt{u_0^2 + (t - t_0)^2 \left(\frac{\theta_E}{\mu_{\text{rel}}} \right)^{-2}}$$

Now let's define a characteristic Einstein ring crossing time

$$(4.13) \qquad t_E \equiv \theta_E / \mu_{\text{rel}},$$

which is simply the time needed for the source to move across the angle θ_E if it has an angular speed μ_{rel}. With this new variable, Equation 4.12 becomes

$$(4.14) \qquad u(t) = \sqrt{u_0^2 + \left(\frac{t - t_0}{t_E} \right)^2}$$

By substituting Equation 4.6 and plugging in some typical values, one can recover a more useful, quantitative form of

t_E as a function of various physical parameters:

$$t_E \approx 35 \text{ days } \left(\frac{M}{M_\odot} \right)^{1/2}$$
$$(4.15) \quad \times \left(\frac{\pi_{\text{rel}}}{125 \ \mu\text{as}} \right)^{1/2} \left(\frac{\mu_{\text{rel}}}{10.5 \ \text{mas yr}^{-1}} \right)^{-1}$$

Thus, a microlensing light curve is parametrized by the time of maximum magnification, t_0, the Einstein ring crossing time, t_E, and the scaled impact parameter, u_0. The first two parameters can be read by eye from the light curve based on the location of maximum light and the width of the light curve, for t_0 and t_E, respectively. The impact parameter can also be estimated based on the level of maximum magnification under the assumption that $A_{\text{max}} = 1/u_0$, where A_{max} is measured with respect to a nonlensed baseline level.

It is important to note that in this simplified picture only two stars are involved: the source and the lens. However, in reality the angular region surrounding the lens–source system contains additional stars that pollute and modify the light curve shape. Rather than measuring a magnification by itself, one measures a flux increase $F_{\text{tot}} = A(t) F_{\text{source}} + F_{\text{BG}}$, where F_{source} is the flux from the source star and F_{BG} is the light from the lens and any other nearby, blended stars. Accounting for blended light is key to properly modeling the physical parameters of a system based on an observed light curve (Di Stefano & Esin, 1995).

Lensing of isolated stars is useful for, e.g., studying the properties of stars on the other side of the Galaxy. The magnification caused by a microlensing event can increase the effective size of one's telescope and open up a view of distant stars normally too faint to study spectroscopically. However, it is when stars and their planets act as lenses that the technique shines as a method of finding and studying exoplanets.

4.3 The Microlensing Signal of a Planet

Now imagine a source star that is perfectly aligned with a lens star, and suppose the lens has a planet at some projected separation $\rho = \theta_E$. The planet will fall right on top of the Einstein ring, and since the planet is massive, it can lens the light in the ring. This will cause the ring to become slightly brighter for the same reason the lensed source becomes brighter.[3] This situation holds for imperfect alignments as well, since the angular separations of the major and minor images, $y_\pm$, are comparable to θ_E. This situation is illustrated in Figure 4.4.

Is it likely that a planet around the lens star will have a projected separation comparable to θ_E? If θ_E corresponds to a separation of 1000 AU, we might not expect planets to act as lenses along with their parent stars. Similarly, if the distance is less than a stellar radius, we shouldn't hold out hope of planetary microlensing signals. Let's do the calculation.

---

[3] However, if the planet crosses the minor image interior to the Einstein ring, it will actually cause the light to *decrease* rather than increase (Mao & Paczynski, 1991).

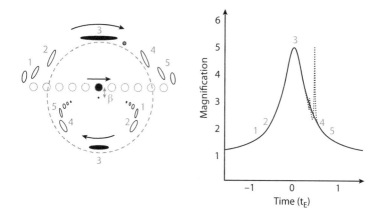

Figure 4.4. A microlensing time series as projected onto the sky. *Left:* The source (gray circles) moves due to its relative proper motion to the lens (small dot at center) with an impact parameter β. The Einstein ring is shown as a gray dashed circle surrounding the lens. The proximity of the source and lens produces two images: the major image exterior to the Einstein ring, and the minor image interior to the ring (black open ovals). As the source moves from left to right, the images move clockwise from position 1 to 5. The lens star has a planet (small gray dot) between positions 3 and 4. *Right:* The resulting light curve is the total flux from the source, lens and images, which produces a magnification relative to the source light, assuming there are no nearby blended light sources. Between positions 3 and 4, the planet's gravitational field acts as a lens, causing an additional magnification (or demagnification). Image adapted from Figure 1 of Gaudi (2011).

At a distance D_L, the small angle $\theta_E \approx R_E/D_L$, where R_E is the physical Einstein ring radius. Using Equation 4.7, θ_E is about 0.71 mas for a 0.5 $M_\odot$ lens at a distance $D_L \approx 4$ kpc and a source at the galactic center

($D_S \approx 8$ kpc), which corresponds to

$$(4.16) \qquad R_E \approx \theta_E D_L = 2.85 \text{ AU}$$

This distance is comparable to the asteroid belt in our Solar System, so it is very reasonable to expect planets around lens stars at a separation comparable to the size of the Einstein ring. I always find it amazing that the scale of the Galaxy and the nature of gravity conspired to result in an Einstein ring radius that is comparable to the semimajor axes of planets! This is one of those happy coincidences I like to remind myself about whenever I feel unlucky because, say, I lose a telescope night to weather.

The planet can be considered as an independent lens with its own value of t_E and θ_E. Since, in general, $t_E \sim M^{1/2}$, the planetary lensing signal will have a much shorter duration than the stellar lensing signal. For example, a Jupiter around a $0.5\,M_\odot$ M dwarf will produce a microlensing light curve that is only 3% as long as the primary signal, or about 1 day. An Earth-mass planet will cause a signal with a duration of only ~ 1 hour.

Also, the mass cannot be solved exactly for the same reason that the lens mass cannot be measured: the exact distances to the lens and source are generally unknown. However, since the star and planet are at the same D_L, their mass ratio, q, can be measured from the ratio of light curve durations:

$$(4.17) \qquad q \equiv \sqrt{\frac{M_P}{M_\star}} = \frac{t_{E,P}}{t_{E,\star}}$$

where $t_{E,\star}$ and $t_{E,P}$ are the Einstein ring crossing times of the star and planet, with masses $M_\star$ and M_P, respectively. Thus, the ratio light-curve widths of the lens star and planet provide a measure of the mass ratio between the planet and star.

Finally, the star–planet separation, s, is encoded in the delay between maxima of the star and planet magnifications. However, this time is measured with respect to $t_{E,\star}$, and provides only the scaled separation between the planet and star, $s \equiv a_p/R_E$, where a_p is the physical separation between the planet and the star projected along the sky plane. Thus, just as with other methods of planet detection, one needs to know the properties of the host star in order to properly characterize the planet.

4.4 Microlensing Surveys

As we saw at the beginning of this chapter, the odds of a microlensing event happening at any given time are about one in a million. This isn't quite as bad as the odds of winning the lottery, but it's pretty unlikely. Just as one could, in principle, buy millions of lottery tickets to increase the chance of winning, the way to counter the unlikeliness of microlensing events is to monitor millions of source stars and wait for a lens star to wander by due to its proper motion.

The best place to find a large density of source stars is toward the galactic center, also known as the galactic bulge. As viewed from Earth, these dense star fields are in the Southern Hemisphere. Two highly successful examples of microlensing surveys are the Optical Gravitational Lensing Experiment (OGLE; Udalski, 2003) and the Microlensing Observations in Astrophysics (MOA; Bond et al., 2001). The original goal of microlensing surveys like MOA and OGLE was to search for dark matter in the form of massive yet cool objects known as MAssive Compact Halo Objects, or MACHOs. While MACHOs were never found in great numbers, many stellar microlensing events were detected.

In the late 1990s and early 2000s, the telescopes used by OGLE and MOA had large fields of view (FOV), but unfortunately detectors were too small to cover much of the available FOV. During the initial planet surveys run by MOA and OGLE, the detectors covered only roughly 0.25 deg^2, resulting in many pointings and a typical cadence of only one to three observations per day. While this cadence was sufficient to detect lensing events that last tens of days, the cadence was too low to cover planetary lensing events that occur on timescales of days to hours. This led to the solution of using an alert and follow-up approach as described by Gould and Loeb (1992).

The first step involved identifying high-magnification events that were in progress by noting the gradual increase in a star's flux levels. Once alerted to a microlensing event in progress, step two involved sending out an alert to a network of small telescopes around the world. Some of these telescopes are professional facilities with 1-meter-class telescopes with CCD imagers, such as those used by the Probing Lensing Anomalies NETwork (PLANET; Albrow et al., 1998). Other telescopes in the network are operated by amateurs with telescopes ranging in size from 0.3 meter to 1-meter in diameter. An example of such a network is the Microlensing Follow-Up Network, or μFUN (pronounced "micro-fun"), which demonstrates one of the many ways in which nonprofessional astronomers have been able to make significant contributions to exoplanetary science (amateur involvement in transit follow-up observing is another example).

The present generation of wide-field microlensing surveys takes advantage of arrays of large-format CCD detectors (Gaudi, 2012). The upgraded OGLE-IV telescope in Chile has a 1.3-meter diameter and a 1.4 deg^2 FOV, and it monitors fields toward the galactic center.[4] The upgraded MOA-II telescope, located in New Zealand, has a 1.8-meter aperture and a 2 deg^2 FOV, and it also targets fields near the galactic center (Hearnshaw et al., 2006). These two upgraded telescopes are joined by the Wise Observatory 1.0-meter telescope in Isreal, which is equipped with a 1 deg^2 FOV imager.[5] These three telescopes are dedicated to microlensing photometry and form a longitudinally distributed network for which it is nighttime at at least one site, weather permitting. Additionally, the large FOVs of the telescopes permit cadences of 10–20 minutes, thereby reducing the need for separate follow-up resources. The current-generation microlensing survey, thanks to advances in detector sizes, allows the alert and follow-up steps to proceed simultaneously, thereby detecting more events per year with uniform data quality.

To date, 18 planets have been detected using the microlensing technique, including two double-planet system.

[4]http://ogle.astrouw.edu.pl/main/tel.html.
[5]http://wise-obs.tau.ac.il/news/exoplanets.html.

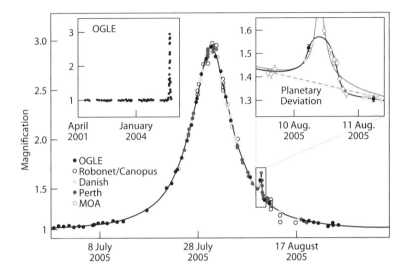

Figure 4.5. The light curve of the microlensing event OGLE-2005-BLG-390. The black closed circles are the nightly observations made by the OGLE survey telescope. Once a rise in the magnification was detected near 23 July 2005, a network of small telescopes began making additional photometric measurements at a much higher cadence. The secondary magnification event caused by a planet lensing the lensed image of the source star is evident near 10 August 2005. The planet responsible for the observed light curve has a mass of 5.5 $M_\oplus$ and it orbits at 2–4 AU from a red dwarf (lens) star. The source is a red giant located near the galactic center (Beaulieu et al., 2006).

An example microlensing planet detection made with the two-step process is shown in Figure 4.5. Notice how the sampling starts off very sparse, with only one or two photometric measurements per night. Then, as the magnification rises, the network of follow-up telescopes is alerted and they start making measurements at a much higher rate, densely sampling the light curve as it rises to maximum magnification. The dense time-sampling continues as the magnification decreases until the secondary magnification of a planet candidate is observed as a blip near August 10. The planetary signal lasts less than a day, but fortunately four different telescopes were able to observe the signal.

The best-fitting model comprises a red dwarf with approximately 20% of the mass of the Sun at a distance $D_L = 6.6 \pm 1.0$ kpc, with a planet orbiting at roughly 3 AU and a mass $M_P = 5.5$ M$_\oplus$. No planets with masses less than that of Neptune have ever been detected beyond 1 AU using any technique other than microlensing, which demonstrates unique parameter space opened up by the microlensing technique.

Present and future microlensing surveys are moving toward dedicated networks of wide-field telescopes distributed around the world. This approach is similar to that of wide-field transit surveys in that having telescopes at a range of longitudes ensures continuous coverage of the target fields. However, the sources monitored by microlensing surveys are much fainter than the targets of transit surveys, and therefore require much larger telescopes. These dedicated telescope networks will make high-cadence observations of their target fields and even low-magnification events (large u_0) will be observed, providing a larger sample of planet detections. By providing sensitivity to even sub-Earth-mass planets at several astronomical units, microlensing surveys will provide vital information about planet formation in a parameter space unreachable by all other current detection techniques.

5

DIRECTLY IMAGING PLANETS

> But what exceeds all wonders, I have discovered four new
> planets and observed their proper and particular motions,
> different among themselves and from the motions of all the
> other stars.
>
> — *Galileo Galilei, Letter to the Tuscan Court,*
> *30 January 1610*

The planet detection techniques presented so far provide
only indirect means of sensing the presence of a planet
around a star. The Doppler, or radial velocity technique
senses the planet's gravitational tug on the host star. The
transit method detects the small decrease in light caused
by the eclipse of a host star by an opaque planet. The
gravitational microlensing method detects the bending of
space-time by the gravitational field of the planet as seen in
the brightening of a background star. Using these detection
techniques, the planet is never seen directly. Instead, the
observed changes in light from a star are used to indirectly
sense the presence of the planet.

This leads to the fourth method that has been used to
detect exoplanets: direct imaging. This technique can be
thought of as taking the picture of a planet next to its
star. This method of planet detection by directly *seeing* a
planet is the oldest discovery technique, dating back to
the discovery of the other planets in the Solar System by

watching the bright Solar System planets move against a background of static stars. However, while the Solar System planets are often the brightest objects in the night sky and therefore easily detected, planets around other stars are far more difficult to directly detect for two primary reasons: they are faint and they are located in close proximity to their extremely bright host stars.

These two related problems are the concepts of *contrast* and *angular resolution*. Contrast refers to the ratio of flux between the star and the planet, and this ratio is typically extremely large. While stars are powered by intense nuclear fusion in their interiors, planets shine either by reflecting a small portion of the star's light or by emitting thermal radiation as a by-product of their formation. Not only are planets faint compared to their host stars; they are also very difficult to discern from stars due to their proximity on the sky. Planets and their stars are separated by distances that are minute compared to the distance between the observer and the star–planet system. Without an instrument that provides high angular resolution, the two sources of light will become blended together, with the planet hidden in the glare of the nearby star. Thus, the challenge of directly imaging a planet comes down the problem of separating an extremely faint image of a planet from the extreme glare of its host star.

5.1 The Problem of Angular Resolution

The first problem is related to how close an planet appears to its star when viewed from far away. The angular separation, θ, between a star a distance d away from the observer and its planet that orbits at a semimajor axis a from its host star is given by $\theta = \sin^{-1}(a/d)$, where a is the semimajor axis of the planet's orbit, assumed to be viewed pole-on for simplicity. As long as the semimajor axis is small compared to the distance to the observer, i.e., $a \ll d$, then $\sin^{-1}(a/d) \approx a/d$. However, the trade-off is that since $d \gg a$, the planet will appear very close to its star. In this context "close" means that a very small angle separates the planet and the star as projected on the sky, so that discerning light from the planet from starlight is very difficult.

The angular resolution of a telescope refers to the ability to discern a small angle between two sources of light on the sky. If the two objects cannot be resolved, then their light will be blended in the image and will appear as a single object. High angular resolution corresponds to the ability to measure small angular separations on the sky. The fundamental limit to angular resolution is set by the size of the telescope aperture (its primary lens or mirror), and larger apertures provide better limiting angular resolution. This limiting angular resolution is also known as the "diffraction limit" of a telescope, given by $\theta_{\min} = 1.22\lambda/D$, where λ is the wavelength at which the observation is made and D is the diameter of the telescope. Note that λ and D must be in the same units.

For observations made at near-infrared wavelengths, say at 1.6 microns on the Keck 10-meter telescope, the limiting angular resolution is $1.22(1.6 \times 10^{-6} \text{ m})/(10 \text{ m}) = 2 \times 10^{-7}$ radian. Radians are useful in trigonometry, but in astronomy small angles are generally given in units known as seconds of arc, or arcseconds. There are 60 arcseconds per arcminute, and 60 arcminutes per degree, so an arcsecond is a tiny fraction (1/3600) of a degree. Observational astronomers usually have the handy conversion factor from radians to arcseconds memorized: there are 206,625 arcseconds per radian, which is approximately 2×10^5 arcseconds/radian. Thus, the diffraction limit of a 10-meter mirror at 1.6 microns is 0.04 arcsecond, or 40 mas.

As a concrete example, Jupiter orbits 5.2 AU from the Sun. Because of the way in which a parsec is defined, there are 206,625 AU in a parsec, just as there are 206,625 arcseconds in a radian.[1] Thus, the angular separation between Jupiter and the Sun as viewed from a distance of 10 pc is $\theta = 5 \text{ AU}/10 \text{ pc} = 500$ mas. At the diffraction limit of the Keck telescope, the separation between our hypothetical planet and star are readily discerned. However, if the planet is in a Jupiter-like orbit around a star 100 pc distant, then the angular separation shrinks down to 0.05 arcsecond, or 50 mas, making it just barely resolved at 1.6 microns with the 10-meter Keck telescope. In principle, such a detection could be made, albeit right at the hairy edge. But in practice there are other complicating factors.

The most important complication for observations made from ground-based telescopes is related to the thin layer of air between the telescope and the star–planet system. The Earth's atmosphere exhibits temperature fluctuations in the form of warm and cool pockets of air that move with the wind above the telescope. When you fly in an airplane, these fluctuations are felt as turbulence.

---

[1] A star at 1 pc will move by 1 arcsecond as the Earth moves by 1 AU. The reader can confirm my statement from here.

This turbulence also affects the images of astronomical sources as viewed from the Earth's surface. This is because air of different temperatures has different indices of refraction, and light is bent in different directions when traversing between warm and cool patches.

This deflection of starlight causes stars to twinkle as their light is deflected into angles in and out of the path to our eyes. Turbulence also causes images of stars to be smeared out over exposure times longer than several seconds. This blurring effect in astronomical images is referred to as "seeing." Nights with very stable air are said to have good seeing, perhaps 0.4 arcsecond at infrared wavelengths. Nights with bad seeing can look perfectly clear to the naked eye, but through a telescope the images of stars can be smeared out by several arcseconds, much to the dismay of observers.

Thus, even under good seeing conditions the Earth's atmosphere imposes the most severe limitation to the attainable angular resolution, setting a limit that can be orders of magnitude worse than the diffraction limit. For example, at Mauna Kea a good night might have 0.4 arcsecond seeing compared to a 0.04 arcsecond diffraction limit of the Keck 10-meter telescope. One solution is to observe from above the Earth's atmosphere by using a space telescope such as the Hubble Space Telescope (HST), or in the future the James Webb Space Telescope (JWST). However, space-telescope facilities are heavily oversubscribed for a wide variety of astronomical studies. HST is a useful tool for directly imaging planets, but observing time is hard to come by. Also, for all its greatness, HST has a relatively small 2.4-meter primary mirror, which limits its attainable resolution.

Fortunately, there is a technological solution that enables high angular resolution from ground-based telescope facilities. If the Earth's atmosphere distorts the light from an astronomical object, then one can deform the telescope's optics to compensate. This technique is known as "adaptive optics," and it is commonly used on the world's largest telescopes in order to obtain diffraction-limited images.

5.1.1 Adaptive Optics

Consider a star at a distance $d \gg R_\star$, such that the surface is not resolved and the star therefore appears to be a point source of light in the sky. The star emits light approximately isotropically (i.e., in all directions), and this light can be thought of as concentric spherical shells representing a series of consecutive wavefronts as light propagates through space. By the time these wavefronts reach the observer far, far away, the radius of curvature is very large, so they appear to be plane-parallel wavefronts with an infinite extent. When these waves (photons) are detected by the telescope, the intensity pattern distributed as a function of angle on the sky is remapped to a spatial location on the instrument's detector. This mapping is equivalent to a Fourier transform from angular frequency on the sky to spatial location on the detector.[2]

---

[2]For the purpose of this book, think of an optical system with a lens as a "Fourier transform machine." Herein I present a heuristic example of Fourier optics for the single purpose of illustrating how an adaptive optics system works. However, a proper treatment of Fourier optics is beyond the scope of this book.

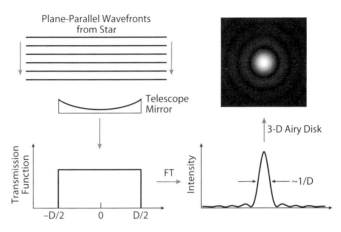

Figure 5.1. A simplified, one-dimensional representation of the optical path of a telescope. Plane-parallel wavefronts from a distant source are incident upon a telescope aperture with width D (*upper left*). The resulting transmission function is a tophat function, with zero transmission outside of the aperture (*lower left*). The imaging optics perform a Fourier transform of the transmission function, resulting in a sinc function, which has a width that scales inversely with the telescope diameter (*lower right*). The two-dimensional equivalent of the sinc is the Airy function shown in the upper right, also known as the imaging system's PSF. The angular distance to the first null gives the diffraction limit of the telescope.

The one-dimensional representation of these plane-parallel wavefronts incident on a telescope mirror with diameter D is shown in the upper left of Figure 5.1. The resulting transmission function is approximately the shape of a tophat, with zero transmission outside of the aperture, and nonzero transmission for positions $\pm D/2$ from the mirror's axis. The optical system performs the equivalent of a Fourier transform of this tophat function,

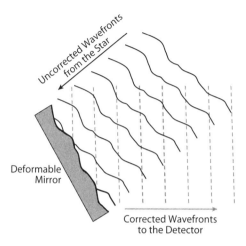

Uncorrected Wavefronts from the Star

Deformable Mirror

Corrected Wavefronts to the Detector

Figure 5.2. Illustration of the action of an adaptive optics system's deformable mirror. The wavefronts from the star have been deformed by temperature fluctuations in the Earth's atmosphere, which would normally lead to a blurry image. The deformable mirror takes on a surface figure that corrects the incoming wavefronts and makes them parallel again. Once the corrected wavefronts are imaged by the camera's detector, the image is much sharper.

resulting in a sinc function of the form $\sin(x/D)/x$. As with all Fourier pairs, the corresponding functions have characteristic widths that scale inversely with one another. The wider the telescope mirror, D, the narrower the sinc function.

The two-dimensional version of the sinc function is the Airy function, shown in the upper right of Figure 5.1. For perfect telescope optics and no atmosphere between the star and telescope, the Airy function is the telescope's *point spread function* (PSF), and the angular distance to the first null is the diffraction limit. However, turbulence in the Earth's atmosphere, as well as imperfections in the telescope and instrument optics, will result in a broader, less ideal PSF. The job of an adaptive optics system is to remove these deleterious effects and recover a diffraction-limited PSF.

The job of an adaptive optics system is to smooth out the corrugated wavefronts that reach the telescope. This is achieved by deforming the telescope mirror in such a way as to remove the "wrinkles" imposed by the Earth's atmosphere. However, rather than changing the giant primary mirror of the telescope, most AO systems use a much smaller deformable mirror in an instrument that is mounted after the telescope focus. The primary mirror is re-imaged onto this smaller deformable mirror where the wavefronts are corrected.

To apply the proper corrections, a part of the light from the star is deflected out of the instrument's light path before it reaches the AO system and is sent to a separate instrument called a wavefront sensor. The shapes of distorted wavefronts are then measured and the necessary corrections are computed and relayed to the deformable mirror. The control loop that senses the wavefronts, and then computes and applies the corrections, operates many times per second, and modern systems can reliably recover diffraction-limited images on 10-meter-class telescopes.

5.2 The Problem of Contrast

The second problem is related to two simple facts: stars are bright and planets are faint. This difference in brightness is commonly referred to as the star–planet contrast, which is a ratio of the flux received from the two objects. There are two ways in which a planet will emit. The first reason a planet will shine is due to the starlight reflected off of the planet's surface. This is the reason that Jupiter is visible in the night sky.

The other source of radiation is the planet's own thermal emission. Planets, like stars, can be thought of as blackbodies due to their nonzero temperatures. Young planets are still contracting from their formation, and therefore emit a large amount of thermal radiation. After planets reach their equilibrium radii, they still retain some of their natal thermal energy, as well as energy received from the star.

To get a sense for the contrast between Jupiter and the Sun, let's first consider the amount of reflected starlight that will be visible from afar. The Sun emits a total power $L_\odot$ approximately isotropically, so that at any distance a away from the Sun, the flux will be the power output divided by the surface of a sphere at a distance a, or $L_\odot/(4\pi a^2)$. The planet has a projected area πR_P^2 and therefore receives an amount of power $L_{\text{receive}} = (1 - A) \times L_\odot/4(R_P/a)^2$, where A is the albedo, or the amount of sunlight that is reflected. For the purpose of this calculation, let's assume $A = 0.3$. The ratio of reflected light from Jupiter compared to the light from the Sun is therefore

$$(5.1) \qquad f_{\text{reflected}} = \frac{1}{4}(1 - A)\left(\frac{R_P}{a}\right)^2$$

or more quantitatively

$$(5.2) \quad f_{\text{reflected}} \approx 1.7 \times 10^{-9} \left(\frac{R_P}{R_{\text{Jup}}} \right)^2 \left(\frac{a}{5 \text{ AU}} \right)^{-2}$$

For the case in which Jupiter's thermal emission is observed, we can approximate both the Sun and Jupiter as blackbodies. The total power output of a blackbody depends only on the object's temperature and radius

$$(5.3) \qquad\qquad L = 4\pi R^2 \sigma T^4$$

where R and T are the stellar radius and temperature, respectively, and σ is the Stefan-Boltzmann constant. The contrast of two different blackbodies, say Star 1 and Star 2, is given by the ratio of their luminosities. This ratio is $L_1 / L_2 = (R_1 / R_2)^2 (T_1 / T_2)^4$. Jupiter has an average surface temperature of 134 Kelvin (a unit of temperature with symbol K) compared to the Sun's 5777 K, and a radius approximately 10 times smaller than the Sun's. The planet–star contrast between the Sun and Jupiter is therefore 3×10^{-9}. The Sun emits 330 million times more power than Jupiter.

There are a couple of ways around this contrast problem. The first is to restrict observations to where the planet is emitting most of its light. While the Sun puts out most of its energy at visible wavelengths close to 0.5 micron, where the human eye is most sensitive, a Jupiter-sized planet will emit most if its energy at infrared wavelengths. Wein's Law provides a handy relationship between the wavelength corresponding to the peak of a blackbody's energy output and its temperature. The wavelength at the maximum of the blackbody function, λ_{max}, measured in microns, is given by

$$(5.4) \qquad \lambda_{max} = \frac{3000 \text{ K}}{T} \ [\mu m]$$

with the temperature, T, measured in Kelvin (K). This relationship predicts that Jupiter, with $T = 134$ K, will emit most of its energy near 22 microns. However, the sun emits relatively little power at such long wavelengths, and the contrast is reduced from 300 million when all wavelengths are considered, to a much more manageable 50,000.[3] Thus, for the purposes of maximizing the contrast between a planet and star, it is advantageous to observe planets at infrared wavelengths, where the star–planet contrast is greatly reduced.

Another way of decreasing the star–planet contrast is to observe young stars that have correspondingly young planets. At roughly 10 million years after its formation, a Jupiter-mass planet will still be much hotter than our own Jupiter, and it will be larger because it will still be experiencing gravitational contraction. At this age, the planet is 3.7 times hotter and 1.3 times larger than Jupiter. As a result, most of its emission will be near 7 microns, and the star–planet contrast is a mere 4000. At the time of this book's writing, all of the confirmed planets directly imaged around hydrogen-burning stars are younger than 100 million years, or less than 2% of the age of the Sun.

[3]This ratio is obtained by integrating the Planck function, B_λ, from 10 to 30 microns at the Sun's temperature.

A technological solution to the problem of contrast is to use an instrument called a *coronagraph* to block out the star's light. This technique is analogous to holding up your hand to shield your eyes from sunlight while driving west during sunset. The coronagraph blocks light from the star but allows light from around the star to pass into the instrument. Coronagraphs can also be designed to reduce the intensity of the Airy rings in a diffraction-limited image. This decreases the star–planet contrast and makes it easier to hunt for planets in the angular region around the star.

However, even when one is looking for hot planets around young stars using adaptive optics and a corona-graph, there remains another obstacle to seeing the light from a planet. This additional problem is related to the imperfections in the optics of the imaging instrument, which lead to quasi-static brightness fluctuations around the image of the star. These fluctuations are known as "speckles" and they are notorious for mimicking the point-source nature of a faint planet. Note that these speckles are distinct from the brightness fluctuations caused by the Earth's atmosphere. While atmosphereic speckles have life-times of tens of milliseconds, speckles due to instrumental optics have much longer lifetimes. Their intensities vary in time as the properties of the instrument change due to temperature variations, or because the instrument weighs hundreds of pounds and is hanging off the back of a telescope, subject to flexure as the telescope moves. The quasi-static nature of speckles means that they can't be averaged out over multiple exposures. This gives rise to a difficult problem of discriminating between speckles and small point sources of light that could be a planet.

One clever technique of distinguishing planets from speckles involves disabling a key function of the telescope, namely the mechanism that corrects for the rotation of the telescope with respect to the sky. To understand how this is done, it is important to keep in mind that the speckles are artifacts of the optical system. An individual speckle can, in principle, be traced to a specific optical element that is fixed with respect to the imaging detector. On the other hand, even though the planet looks like a speckle, its image can be traced back to a specific angular position on the sky rather than the optics. A way to distinguish between a speckle and an astrophysical light source is to take a sequence of images while allowing the instrument to rotate with respect to the sky. The speckles will appear to remain fixed, while the planet candidate will appear to rotate around the target star.

This method of angular differential imaging (ADI) was originally used with the Hubble Space Telescope, which can roll around its optical axis while floating in space in order to change the mapping of angular position on the sky to different pixels on the imaging detector. To use this technique on a ground-based telescope, an astronomer named Christian Marios and his collaborators cleverly noted that telescopes like Keck have an image de-rotator that normally keeps angles on the sky mapped to a fixed position on the detector. By disabling the telescope's de-rotator, Marios and his team found that they could simulate the rolling of a space telescope.

Figure 5.3 shows an example of an image of a star and a low-mass companion. The left panel shows the effects of speckles: they look like point sources, just like the companion. In the right-hand panel, the speckles have

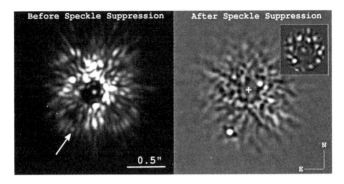

Figure 5.3. *Left:* Image of the Sun-like star HR 7672 made with an adaptive optics system and a coronagraph, without noise suppression. The brightness fluctuations are known as speckles and are the result of imperfections in the telescope and instrument optics, as well as residual wavefront errors that are not adequately corrected by the AO system. The faint companion is a brown dwarf (HR 7672 B) that orbits at a distance of 18 AU from the central star, and is indicated by an arrow. Note that the companion is fainter than many of the speckles. *Right:* The same image after aggressive speckle suppression. The companion now stands out clearly compared to the residual noise. The inset image shows the inner region of the image near the coronagraphic mask, demonstrating a nearly diffraction-limited image of the star. Figure from Crepp et al. (2012)

been suppressed by using ADI, and the faint companion stands out clearly.

Another consequence of the speckles being related to the optics of the instrument is that their locations with respect to the central star will vary with wavelength. Thus, speckles can be separated from astrophysical sources by taking a sequence of images at different wavelengths. While the PSF of an astrophysical source will remain at a constant position when observed at different wavelengths, speckles will move radially outward with increasing wavelength. This is most effectively accomplished using an instrument known as an integral-field unit (IFU). A grid of tiny lenslets is placed in the light path, and the light is then focused onto a prism or dispersing grating. This results in a spectrum at each point in the grid. These spectroscopic pixels (spaxels) can be divided into a sequence of image pixels at different wavelengths. A major advantage of this technique of spectral differential imaging (SDI) is that one obtains a spectrum of the planet at the same time that it is discovered, thereby allowing instant characterization of the newly discovered planet.

5.3 The Problem of Chance Alignment

Similar to the adage "All that glitters is not gold," all faint point sources near your target star are not planets. In fact, more often than not the faint speck next to a star in an adaptive optics (AO) image is either a speckle or a background star. So even after employing various clever workarounds to decrease the effects of optical aberrations, there are many ways to be fooled when attempting to directly image a planet.

The primary means of verifying an imaged planet involves waiting some amount of time, typically of order a year or two, to see if the planet travels with the star across the sky. Stars in the Galaxy are not static. Instead, stars orbit the galactic center while gravitationally jostling each other. The result is that stars move subtly across the sky. This angular motion is known as *proper motion*, and it is typically measured in milliarcseconds (mas) per year.

Stars nearer to the Earth tend to have larger proper motion than those that are far away, so background stars will appear to remain roughly in place while the target star drifts across the sky. Planets, on the other hand, will be gravitationally bound to the target star and will track the star's motion over the years. For this reason, planet hunters request telescope time initially to conduct their survey and build up a list of candidates, and additional time for follow-up observations for proper-motion confirmation.

5.4 Measuring the Properties of an Imaged Planet

Once a tiny point source of light has been imaged next to a bright star and determined to be gravitationally bound, how does one determine the mass and orbital properties of the object? The two properties that can in principle be measured are the planet's semimajor axis, a, and its mass. The semimajor axis can be estimated by measuring the angular separation, ρ, between the star and its planet. If the star's distance, d, is known, then the projected semimajor axis is given by $a_{\mathrm{proj}} = d\rho$, under the assumption that $d \gg a$ and $\sin \theta \approx \theta$.

Unfortunately, determining the mass of an imaged, low-mass companion is necessarily a difficult, model-dependent process. Present-day direct imaging surveys depend on detecting the thermal emission of planets, and this emission is at its brightest when the planet is young. This is because the thermal energy is generated from the slow, gravitational collapse of the newly formed object. As it gets older, it contracts further and becomes cooler.

The changing temperature of a newly formed planet also results in a change in its spectroscopic properties.

Theoretical models of the interior structures and atmospheric properties of planets are often the only means of relating their observed properties to their physical properties, namely their mass. In addition to the observed properties of the planet, one must also know its age and chemical composition. In practice, astronomers assume that the planet shares the age and chemical makeup of its host star. While measuring the chemical composition of a star is fairly straightforward, measuring stellar ages is extremely difficult. If the star belongs to an open cluster, one can assume the star has the age of the cluster, as determined from other means. However, if the star is isolated, ages can be uncertain by 50–100%.

Despite these limitations, direct imaging provides a means of studying young planets in wide orbits similar to the Solar System gas giants, and beyond. The detection of these young, long-period planets opens up many unique opportunities to study the atmospheric properties of exoplanets by measuring the spectrum of light emitted directly from the planet's photosphere. These direct spectroscopic observations of planetary atmospheres will be used to test and calibrate models of planetary interiors. Further, the chemical composition of the planet can be used to infer the details of its formation, such as where in the protoplanetary disk it formed. Finally, additional images of the planet taken over timescales of years will allow astronomers to measure the orbital properties of the planet, which can be compared to the large sample of Doppler-detected planets.

6

THE FUTURE OF PLANET HUNTING

Despite being one of the youngest areas of study in astronomy, exoplanetary science has been extraordinarily successful and the field is still growing rapidly. The first exoplanet securely detected around a hydrogen-burning star was discovered 1996 (Mayor & Queloz, 1995; Marcy & Butler, 1996). In the years since, the number of confirmed exoplanets has exceeded 1500, and the number of candidates discovered by the NASA *Kepler* Mission numbers in the thousands.[1] Once the initial technological hurdles were cleared by the first planet hunters, the rate of discovery increased exponentially each year.

The question posed in the title of this book, "How do you find an exoplanet?" is premised on the notion that finding planets outside our Solar System remains a worthwhile endeavor. However, given the large collection of known exoplanets already in hand, it is reasonable to question whether the discovery of additional exoplanets is warranted. Have the goals of planet hunters already been met? To address this question, it is important to review the

---

[1] Estimates place the fidelity of the *Kepler* candidates at about 90%, with only 10% likely to be false positives.

primary goals of this young field of study. To my mind, the goals are threefold: gaining a broader, Galaxy-wide perspective on planetary systems; understanding the origins of the Solar System through knowledge of planet formation in general; and searching for life outside our planet.

6.1 Placing the Solar System in Context

The first goal of exoplanetary science is to place our Solar System, with its eight planets and yellow dwarf central star, into a broader, galactic context. This goal follows directly from the results of the Copernican Revolution. We know that the Earth and the system of planets in which it resides does not represent a special place in the Universe. The other stars in the night sky are like the Sun in that they are approximately spherical collections of gas powered by hydrogen fusion in the cores, and as such they share similar formation and evolutionary histories. Each formed from the collapse of a large cloud of dust and gas resulting in a central concentration of material that becomes the star. The remaining infalling material formed a flattened distribution known as a protoplanetary disk, and planets presumably formed out of the disk. We see young stars forming throughout the Galaxy, and we occasionally even observe their protoplanetary disks. Today, we can observe other planetary systems, and these systems often have architectures (semimajor axes, planet masses, eccentricities) that differ significantly from that of the Solar System.

The existence of exoplanets demonstrates that the Solar System is not the lone planetary system in the Galaxy.

While few scientists would have assumed that there existed only a single planetary system, we have known of multitudes of systems for only the past two decades. That the Solar System has analogs elsewhere is, relatively speaking, brand new information not known to previous generations of humans. Thus, in a broad sense we have accomplished the goal of attaining a broader perspective. However, when compared with the Solar System, the details of exoplanetary systems are surprising in many aspects—a direct result of how limited our view of the Cosmos was, and in many ways still remains.

When planet hunters first set out to find new planets around other stars, the Solar System served as their primary guide for what to expect. If the Sun's radial velocity were monitored from several parsecs away, the most readily detectable planet would be Jupiter, which induces a Doppler wobble of approximately 12 m s^{-1} if viewed edge-on, with a period of approximately 12 years. Detecting Jupiter analogs was thought to be a long-term process requiring dedication over the course of decades. However, every one of the first exoplanets defied this expectation.

The first Jupiter-mass exoplanet discovery around a normal, hydrogen-burning star is usually attributed to Michele Mayor and Didier Queloz, who announced a planet in a 4.2-day orbit around 51 Pegasi in 1996 (the planet was confirmed in short order later that year by Geoff Marcy and Paul Butler). This hot Jupiter challenged our understanding of how gas-giant planets form and how their orbits subsequently evolve because planets as large as Jupiter likely cannot form from the hot, rarefied gas right next to a young star. As a result, the hot, giant planet 51 Peg b must have *migrated* inward from its birthplace, which was most likely at several astronomical units from the central star. Thus, this single discovery greatly broadened astronomers' perspective on the variety of possible planetary system architectures.

While 51 Peg b is traditionally referred to as the first exoplanet around a normal star, several planets were discovered in the years preceding 1996. One example is the system of three small planets discovered in 1992 by Aleksander Wolszczan and Dale Frail orbiting a pulsar named PSR B1257+12 (Wolszczan & Frail, 1992). Pulsars are the skeletal remains of stars more massive than the Sun and are produced by supernova explosions. Compared to hydrogen-burning stars, pulsars—more generally known as neutron stars—are extremely strange objects. They have masses comparable to the Sun's, yet they are extraordinarily small and therefore very dense. Neutron stars have diameters the size of a small city and they emit magnetically channeled beams of radiation from their poles while they rapidly spin. If viewed with the correct geometry, the beam sweeps through the observer's line of sight, resulting in a lighthouse phenomenon with regular pulses of light observed at regular intervals.

Wolszczan and Frail noticed that regular periodicity of one of their pulsar targets underwent modulations caused by the gravitational pull of three planets, and they were able to measure the masses and orbital properties of three low-mass planets in a manner analogous to the Doppler method (Wolszczan & Frail, 1992; Wolszczan, 1995). However, it is unlikely that these "pulsar planets" orbited the massive progenitor star before it exploded in a supernova that formed the neutron star since the planets likely would not have been able to survive the blast. Instead, the pulsar planets are widely believed to have formed from the debris left over from the supernova explosion, and as a result they are not generally thought of as analogs to the Solar System planets. Nevertheless, these strange planets and their dead central star most certainly broadened humanity's perspective on the nature of planetary systems.

Most people are surprised that the discovery of exoplanets around hydrogen-burning stars dates back even further than 51 Peg b and the pulsar planets. In the late 1980s, Harvard-Smithsonian astronomer David Latham and his collaborator Tsvi Mazeh, from the Tel Aviv University in Israel, were conducting a radial velocity survey of bright stars using a new instrument called the CfA Digital Speedometer. One of their stars, HD 114762, exhibited oscillations in its radial velocity with a period of roughly 83 days and an amplitude and shape consistent with a minimum mass of 11 M_{Jup} and an eccentricity of 0.34. In the title of the paper announcing the detection, Latham et al. (1989) cautiously called the orbital companion a "probable brown dwarf," because an orbital inclination $\sin i < 0.85$ would result in a true mass $m_p > 13 M_{Jup}$, making the object too massive to be considered a planet. Additionally, the object was an order of magnitude more massive than the most massive Solar System planet, which seemed too strange to be considered analogous to anything in the Solar System.

However, in the decades since the discovery of HD 114762 b it has been firmly established that brown dwarfs with masses in excess of $13\,M_{Jup}$ are much rarer than planets. Also, while planets more massive than Jupiter are fairly rare compared to less massive planets, planets with masses in the range $1\text{--}13\,M_{Jup}$ do exist. Thus Latham et al.'s discovery likely represents the first exoplanet detected around a Sun-like star. At a celebration of Dave Latham's 50 years in astronomy, the community enthusiastically decided to name the object "Latham's Planet."[2]

This moniker is a tribute to the hard work of the early planet hunters, who toiled without the certainty that they would find anything. It is also a clear demonstration of how our view of planets, and indeed even our definition of what counts as a planet, has been greatly modified and enhanced by our newfound galactic perspective. Latham's Planet demonstrates that giant planets can have masses ranging up to an order of magnitude greater than that of Jupiter and they can have eccentricities that are an order of magnitude larger than the eccentricities of the Solar System planets. Further, the orbital periods of Jupiter-mass planets can be as short as a few days.

These early discoveries brought with them surprises that challenged our myopic expectations of planetary systems that were based on a sample of one. In the decades since the original exoplanet detections have continued to surprise astronomers. Indeed, the new normal for planets is that they consistently challenge prior assumptions and expectations. As new instrumentation becomes more sensitive and the time baselines of existing surveys increase, the

---

[2]Whether the International Astronomical Union agrees with this naming scheme is another question and Dave himself insists that the planet already has a perfectly good name.

surprises keep on coming and show no sign of abating anytime soon. Planets have thus far been found around an enormous variety of stars, ranging from tiny red dwarfs with 15% of the Sun's mass $(0.15 \, M_\odot)$, all the way up to red giant stars with masses in excess of $3 \, M_\odot$. Remnants of stars have planets, as do binary stars with planets orbiting one or even both components of double-star systems. As the search for planets continues, new types of planetary systems are revealed, and these revelations continue to shape our notion of what constitutes a "typical" planetary system in the Milky Way.

6.2 Learning How Planets Form

The second goal of searching for exoplanets is to use the observed distribution of exoplanetary system architectures and the rate at which planets are found around other stars as clues about the planet-formation process (or processes). The physical mechanisms that result in the assembly of relatively huge planets from the microscopic molecules and dust grains in a protoplanetary disk remain largely mysterious to astronomers because of two main factors. First, young stars and their protoplanetary disks reside in star-formation regions that are often shrouded by gas and dust, making the planet-formation epoch, and the important processes in action during this time, difficult to observe. Second, the planet-formation process is extremely short relative to the lifetime of a typical star. Censuses of disks in star-formation regions demonstrate that protoplanetary disks, and hence the epoch of gas-giant formation, last only 3 million to 10 million years, and terrestrial planet formation lasts until about 100 million years.

On a human timescale, tens to hundreds of millions of years represents an eternity. However, 10 million years represents only 0.2% of the Sun's main-sequence lifetime. To put it in numbers that are easier to think about, if the entire lifetime of the Solar System were condensed into a single day, and right now is 11:59 p.m., then the epoch of planet formation occurred this morning from midnight to about 12:02 a.m. for Jupiter and Saturn. The relatively long period during which the Earth formed lasted until about 12:30 a.m. The challenge faced by exoplanetary scientists is to reconstruct what happened during that brief period "this morning" based on clues found much later in the evening.

The problem of limited visibility will soon change as the Atacama Large Millimeter Array (ALMA) comes online in the next few years. ALMA is a huge collection of radio telescopes in a high desert plateau in Chile. The telescopes are spread out over a large area and they work together to form a telescope aperture that has an effective diameter that spans the largest separation between two of the telescopes. The result is a telescope, known as an *interferometer*, that has unprecedented angular resolution and light-gathering ability (sensitivity). ALMA will be able to peer through the enshrouding material around young stars in order to catch planets in the process of formation. The downside is that only a small number of nearby, young star-forming regions are accessible for study, which will ultimately limit the statistical utility of ALMA's observations of planets in the process of formation.

By observing exoplanetary systems around mature stars that reside outside of star-forming regions astronomers can gather much larger sample sizes, with the trade-off that these systems are well past the time of formation. However, present-day planetary systems represent the end states of the mechanisms of planet formation and evolution. This collection of known systems is the product of the hard work of planet hunters, who are, generally speaking, observational astronomers. In astronomy there is often a close relationship between observers and theorists. In some fields, theorists devise hypotheses that motivate future observations. However, in the field of exoplanetary science the process works primarily in the other direction, with observers discovering previously unimaginable planets and orbital architectures, which the theorists attempt to explain using sophisticated planet-formation models. In the study of exoplanets, observers drive the field by discovering the boundary conditions that the theorists attempt to reproduce with their models.

Ultimately, a better understanding of how planets form and evolve cycles back and informs our own origins. As outlined in the previous section, gaining a broader perspective of planetary systems throughout the Galaxy helps us complete the Copernican Revolution, and the large numbers and rich variety of known exoplanets defin-itively demonstrate that the Solar System's existence is not particularly special. The continued discovery of additional exoplanets builds our statistical sample, providing a clearer picture of the possible outcomes of the planet-formation process, which theorists attempt to reproduce. The model that ultimately provides the best match to the various observed systems will represent an excellent candidate for a *general* description of planet formation, which produced the Solar System as one of the many possible outcomes. This would bring about the completion of the Copernican Revolution, as described in the introduction of this book.

However, while scientists are motivated to see the Solar System as just one of many planetary systems, the Earth will always feel special simply because it is *our* planet, *our* home. As a result we have a vested interest in understanding its specific history. The origin story of our planet is also the story of our origin, and learning about our history is compelling on both a scientific and emotional level. By studying exoplanets as the end states of the planet-formation process, we are ultimately looking into our own past.

Put another way, we can use evolutionary biology to examine the relatively recent origin and history of the diversity of life on Earth. Looking much further back, cosmology is the study of the origin of the entire Universe. Exoplanetary science fits nicely in between.

6.3 Finding Life Outside the Solar System

My earliest conception of exoplanets took the form of the fictitious worlds in the original *Star Wars* movies. There was the ice world Hoth; the arid desert terrain and double suns over Tatooine; the dank swamps of Degobah; and the dense, lush forests of Endor. Yet, despite the name of the series, no close-up views of stars appear in any of the original three *Star Wars* movies. Stars form the background for duels among spaceships, there's the iconic double sunset on Tatooine, and stars streak by when the Millennium Falcon makes the jump to light speed. But the richest visual tapestries in the movie trilogy are reserved for the surfaces of various planets.

A reason for this feature of most science fiction, and why planets hold such a special interest in the minds of astronomers and the general public alike, was summed up nicely in my classmate Jason Wright's PhD thesis:

> Of all of the topics of study in astronomy, exoplanets hold a special place in the imagination. More than stars, nebulae, or galaxies, they are *places*; both because of their similarities to this place, the Earth, and because in the popular imaginings of science fiction they are so often *destinations*. As other examples of our home, they promise clues to the origin of life on Earth, and to the existence of life elsewhere.

The scientific reasons of gaining a broader perspective and understanding of planet formation are very compelling. However, for me the most compelling reason for studying planets is because I have always had a vivid imagination for what it would be like to leave the gravitational bounds of the Earth and travel to another *place* in the Galaxy. Even if it were right next door, a mere parsec or so to the α Centauri system, I can imagine no greater feat of science, and no more worthwhile journey for humans, than to travel to another planet around a distant and alien sun.

But as engaging as it is to daydream about space travel and exploring the surfaces of new planets, my sci-fi-fueled imagination must be reigned in by reality. We're not likely to make such a journey anytime soon, and almost certainly not in my lifetime. Yet I can't help but think that humans will make the trip eventually, provided we can make our existence on our own planet sustainable long enough to embark on such an ambitious endeavor.

While space travel is out of the picture for now, the search for extraterrestrial life is under way here on Earth. Many people are familiar with the Search for Extra-Terrestrial Intelligence (SETI), which was featured in the movie *Contact*. The main character, Ellie Arroway (played by Jodi Foster), is based on the real-life astronomer Jill Tarter, who is the former chair of the SETI Institute near San Jose, California. SETI uses various astronomical telescopes to "listen" for electromagnetic emission from nearby extraterrestrial intelligent civilizations that are either intentionally or unintentionally beamed to Earth (Tarter, 2001; Penny, 2011).

One disadvantage of the SETI approach is that it depends on the existence of a very specific type of extraterrestrial life: intelligent life forms that are technologically advanced enough to communicate via electromagnetic signals. However, the emergence of intelligent, communicating life on Earth was very recent and extremely brief compared to the approximately 3.5 billion-year history of life in general on our planet. Additional avenues of searching for extraterrestrial life are opened by relaxing the requirement that life be intelligent and capable of communicating with other planets.

One method is to search for evidence of life by carefully examining the atmospheres of planets using spectroscopy.

Lifeforms interact with and can potentially alter the atmosphere of their planet in ways that are distinct from geochemical processes (Seager et al., 2009). However, detecting this biochemical alteration—also known as the search for biosignatures—first requires the identification of potentially inhabited planets, with rocky surfaces, and life-sustaining atmospheres and temperatures. Measuring the fraction of stars with habitable planets was the primary objective of the NASA *Kepler* Mission (Basri et al., 2005), and future efforts will use these statistics to search for these habitable planetary systems around nearby stars and study their atmospheres for evidence of life. Thus, the search for life requires the discovery of the exact locations in the Galaxy—the astrophysical places—where life exists.

6.4 Giant Planets as the Tip of the Iceberg

The goals enumerated in the previous sections highlight a key progression in exoplanetary science. First, early discoveries of a particular class of planets immediately provide context for the Solar System. As additional planets are discovered and the sample size grows, key correlations are uncovered that provide clues about the planet-formation process and guide the search for additional planets. Eventually, the sample size grows large enough that rare and exciting systems eventually emerge. These rarities sometimes offer the opportunity to study planets around other stars to a level of detail comparable to our knowledge of Solar System bodies.

An example of a well-studied subgroup of planets is the close-in giant planets, including hot Jupiters and other gas giants orbiting much closer to their stars than Jupiter orbits the Sun. As the collection of close-in giant planets grew, one of the first and clearest statistical results to emerge was the so-called planet–metallicity correlation (Gonzalez, 1997; Santos et al., 2004; Fischer & Valenti, 2005). The metallicity of a star refers to the abundance of elements heavier than helium in a star's atmosphere, and stars with metallicities higher than the Sun's value have a much higher probability of having a giant planet than do stars with low metallicities.

From a planet-formation standpoint, this provides an important clue about how giant planets are born. Young stars and their disks share very similar compositions since they formed out of the same molecular cloud, and while protoplanetary disks are short-lived, the metal content of those disks is preserved in the atmospheres of the stars we see today. Stars with high metallicities likely had higher amounts of dust-forming heavy elements, which form the building blocks of the solid cores of giant planets. This important observational result provides another key touchstone for planet-formation models, as any successful theory must explain the planet–metallicity correlation.

In addition to providing clues about the process of planet formation, the planet–metallicity correlation also pointed the way toward finding additional planets: if it is giant planets you seek, then search around metal-rich stars. Two different planet searches led by Debra Fischer and Ronaldo da Silva, respectively, targeted metal-rich stars to increase the number of known hot-Jupiters, with a particular focus on those that transit. Some of the brightest transiting hot Jupiter systems were discovered by these surveys, including HD 189733 b and HD 149026 b (Bouchy et al., 2005; Sato et al., 2005), which have been followed up extensively by ground- and space-based follow-up studies of exoplanet atmospheres (e.g., Knutson et al., 2009; Pont et al., 2013).

The planet–metallicity correlation is an example of how the discovery of an initial collection of planets, in this case the hot Jupiters, led to additional efforts to find and discover ever more planets, eventually leading to rare and exciting examples that can be studied in even finer detail to understand the nature of planets outside the Solar System. A more recent result that will guide future planet search efforts is the remarkable finding that planets smaller than Neptune are much more common than giant planets like Jupiter. This statistical result was initially evident from the population of planets found by Doppler surveys, and has been corroborated by microlensing and transit surveys.

Figure 6.1 shows the distribution of planet radii (in Earth units, $R_{\oplus}$) of planets found by the NASA *Kepler* transit survey. The distribution shows that small planets with radii in the range 1–2 $R_{\oplus}$ are an order of magnitude more abundant than giant planets ($R_P > 4\ R_{\oplus}$) throughout the Galaxy. While this distribution is for orbital periods less than 150 days, the distribution of planet sizes is consistent with the distribution found from Doppler and microlensing surveys. Small planets are much more common than large planets throughout the Galaxy, which is very promising for humanity's hunt for planets like our own and the search for life outside the Solar System.

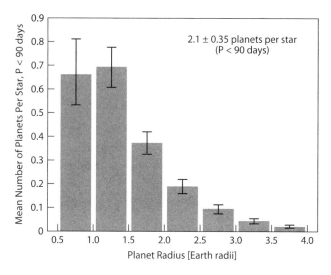

Figure 6.1. The distribution of planet size (radius, in Earth units) for planets discovered by the NASA *Kepler* Mission orbiting M-type dwarf stars (red dwarfs), with periods less than 150 days. This figure is adapted from the study of Morton & Swift (2014), who find that there is on average about two planets per M dwarf in this period range. Since M dwarfs comprise 7 out of 10 hydrogen-burning stars throughout the Galaxy, these types of planetary systems are the most numerous in the Milky Way, and the peak of the distribution near the size of the Earth is promising for future searches of other habitable planets like our own.

From the standpoint of placing our Solar System in perspective with the Galaxy-wide population of planets, the existence of Jupiter and Saturn looks fairly unusual, while our system of inner terrestrial planets looks rather familiar. From a formation standpoint, any successful planet-formation model must simultaneously reproduce the ubiquity of small planets and the rarity of gas giants, while still occasionally producing systems like our own with an admixture of both. Finally, in terms of finding more planets, improving the sensitivity of astronomical instrumentation is key. For example, every fraction of a meter-per-second improvement in the radial velocity precision of future spectrometers will allow planet hunters to climb the steeply increasing mass distribution to find not only smaller planets than have been found so far, but many more of them. Sensitivity to ever smaller and longer-period planets not only moves us closer to finding exoplanets that resemble our own, but, equally important, will reveal the bulk of the planet population in the Galaxy.

In this final chapter I provide several examples of future planet-hunting instrumentation for each of the four detection methods reviewed to this point. This list is not intended to be comprehensive, as the number of instruments either being built or in the planning stage is far too large to be covered adequately in this book, and the list of future instrumentation will no doubt change before the book is even printed. However, with the examples I have chosen I aim to provide a general sense for the direction the field is heading in the near future.

6.5 The Future of the Doppler Method: Moving to Dedicated Instrumentation

I find it remarkable that many of the exoplanets detected before the launch of the *Kepler* Mission were discovered with astronomical instruments and equipment that were built for other purposes. For example, many of the planets detected using the Doppler technique were observed with spectrometers like the HIgh Resolution Echelle Spectrograph (HIRES) at the Keck Observatory (Vogt et al., 1994). HIRES is a multipurpose instrument with many settings to allow astronomers to observe objects ranging from isolated stars to entire galaxies at the edge of the visible universe. The instrument has many adjustable parts that allow for different observing modes. But this flexibility works against the stability needed to measure stellar radial velocities to a precision of meters per second. Highly precise radial velocity measurements with HIRES have been made possible only by affixing "aftermarket" parts, namely the iodine absorption cell that is used to track and correct changes in the instrument. The same holds for many of the precursor instruments to HIRES, including the Hamilton Spectrometer at Lick Observatory (Fischer et al., 2014), the High–Resolution Spectrograph on the Hobby–Eberly Telescope in Texas (Tull, 1998), and the original planet survey using a gas absorption cell performed by Campbell & Walker (1979), who used highly poisonous hydrogen-flouride gas cells a decade before the use of iodine.

The HIRES instrument stands in stark contrast to the instrument used by many European planet hunters, the High Accuracy Radial-velocity Planet Searcher (HARPS) spectrometer at the Paranal Observatory in La Silla, Chile (Mayor et al., 2003) and the HARPS-North instrument at La Palma in the Canary Islands (Cosentino et al., 2012). As their name suggests, the two HARPS spectrometers were built with one purpose in mind: attaining meter-per-second radial velocity precision to find planets.

Unlike HIRES, HARPS is housed in an isolated, underground room beneath the telescope. The room is environmentally controlled, and the spectrometer sits within a large vacuum chamber. Instead of using an iodine absorption cell to set the wavelength scale (mapping of wavelength to pixel in the spectrometer's detector), HARPS uses a thorium-argon emission lamp to place lines of known wavelength next to each stellar spectrum. As a result of its extreme stability, HARPS can achieve a level of RV precision that is roughly a factor of 1.5 to 2.0 better that of HIRES. While HIRES has more planet detections overall, most of the low-mass "super Earths" with minimum masses ($M_p \sin i$) less than Neptune's yet larger than the Earth's have been detected by HARPS. The HARPS spectrometers demonstrate that extremely high precision requires customized, purpose-built instrumentation.

The archetype of the next generation of customized, highy stabilized spectrometers that is planned for the near future is ESPRESSO, which will be mounted on the four-telescope Very Large Telescope (VLT) array (Pepe et al., 2014). While the HARPS instruments are mounted on telescopes with relatively small apertures—the European 3.6-meter ESO telescope at La Silla Observatory in Chile, and the Italian 3.58-meter Telescopio Nazionale Galileo in the Canary Islands—ESPRESSO will harness the extreme light-gathering capabilities of the four 8.2-meter VLT telescopes. It can be mounted to one telescope at a time, or fiber-optic cables can feed the spectrometer from all four 8.2-meter telescopes simultaneously. The extremely high spectral resolution, instrumental stability and advanced optical design enable a target precision of $10 \, \mathrm{cm \, s^{-1}}$ on bright stars, or high-precision radial velocity monitoring of stars much fainter than have previously been observed with even the largest 10-meter-class telescopes of today.

Whereas the ESPRESSO spectrometer will aim for extreme single-measurement precision, moving the state of the art from $1 \, \mathrm{m \, s^{-1}}$ to $10 \, \mathrm{cm \, s^{-1}}$ to enable the detection of Earth-sized planets, the Automated Planet Finder (APF) at Lick Observatory will press forward in the time domain, using high-cadence observations of bright stars (Vogt et al., 2014). As its name implies, the APF will be a robotic telescope that will observe dedicated planet-search target lists on a nightly basis. A similar approach will be taken by the future MINiature Exoplanet Radial Velocity Array (MINERVA) at Mt. Hopkins in Arizona (Swift et al., 2014). Both MINERVA and the APF will trade aperture—1.4-meter and 2.4-meter telescopes, respectively—for high cadence in the search for low-mass planets.

In Chapter 2 we saw that an Earth-like planet around a Sun-like star induces a Doppler amplitude of only about $10 \, \mathrm{cm \, s^{-1}}$. To date, the smallest amplitude detected is the planet candidate around α Cen B, which has $K = 50 \, \mathrm{cm \, s^{-1}}$ (Dumusque et al., 2012). However, it is important to note that one does not need to reach a precision of $10 \, \mathrm{cm \, s^{-1}}$ *per observation* to detect an amplitude $K \approx 10 \, \mathrm{cm \, s^{-1}}$. If the uncertainty in each measurement is σ_0, then the precision in the Doppler amplitude is $\sigma_K \sim \sigma_0 / \sqrt{N_{\mathrm{obs}}}$, where the N_{obs} is the number of observations. By collecting dozens, if not hundreds of radial velocity observations of targets throughout the year, the APF and MINERVA telescope/spectrometers will be able to detect planets with Doppler amplitudes well below its $1 \, \mathrm{m \, s^{-1}}$ single-measurement precision.

Most existing spectrometers built for high-precision Doppler measurements operate at visible wavelengths. This wavelength choice is driven by several factors. First, stars like the Sun put out most of their energy at visible wavelengths between 500 and 600 nm. Second, silicon detectors such as CCDs have their peak sensitivity in this wavelength regime. Also, silicon detectors are much less expensive than the more exotic materials that comprise infrared detectors, such as mercury cadmium telluride (HgCdTe). Finally, the iodine cell provides the highest density of deep absorption features, useful for wavelength calibration, in the interval 500–620 nm. However, Earths in the habitable zones of Sun-like stars have much longer orbital periods and smaller Doppler amplitudes than planets in the habitable zones of less massive, redder M dwarf stars (see Equation 2.24). The challenge is to build spectrometers that can measure precise radial velocities at wavelengths where M dwarfs are bright, namely in the infrared.

Several stabilized, infrared spectrometers are planned for operation in the near future. The Habitable-zone Planet Finder (HPF) on the 10-meter HET will use a HARPS-like approach to extreme instrumental stability, with a vacuum enclosure, temperature stabilization, an emission lamp wavelength calibrator and fiber-optic light feed (Mahadevan et al., 2012). The HPF is specifically designed to search for Earth-sized planets around the least massive M dwarfs in the solar neighborhood.

Two European efforts will also target low-mass stars by observing in the infrared. The Calar Alto high-Resolution search for M dwarfs with Exo-earths with Near-infrared and optical Echelle Spectrographs, or CARMENES, will actually house two spectrometers, one operating in the traditional optical regime and one in the infrared (Quirrenbach et al., 2014). The CARMENES survey will target about 300 nearby M dwarfs using 600 nights of dedicated time on the 3.5-meter telescope at the Calar Alto Observatory. A similar instrument to be mounted on the 3.6-meter Canada France Hawaii Telescope (CFHT) in Hawaii is *un SpectroPolarimÃltre Infra-Rouge* (SPIROU; translated to "A Near-Infrared Spectropolarimeter") and will allow observations out to 2.5 μm (Artigau et al., 2014).

6.6 The Future of Transit Surveys

The success of a transit survey is entirely dependent on two factors: photometric precision and time coverage. The minimum transit depth detectable depends on the photometric precision; smaller transit dips can be seen with less photometric scatter. And since transit events are short compared to the orbital period, continuous, high-cadence coverage is required. For example, the fraction of an orbital period that the planet spends in front of a Sun-like star is the transit duration divided by the period, or T/P. Using Equation 3.4 along with Newton's version of Kepler's Third Law, this fraction is

$$(6.1) \quad \frac{T}{P} \approx 1\% \left(\frac{P}{10 \text{ days}} \right)^{-2/3} \left(\frac{M_\star}{M_\odot} \right)^{2/3}$$

assuming equatorial transits ($b = 0$) and $R_\star = M_\star$ in solar units for main-sequence stars. Thus, even for short-period planets, a star spends most of its time uneclipsed by the planet, and without continuous monitoring, a transit event is easy to miss.

The requirements for high photometric precision and continuous-time coverage argue for a space-based observatory, rather than a ground-based effort. The NASA *Kepler* Mission is an excellent example of a purpose-built space platform for transit survey work. The downside of the *Kepler* approach is its fixed field of view that, while wide for a single telescope pointing, represents only a small fraction of the entire sky. Since bright stars are much more sparsely distributed over the sky than faint stars, the vast majority of the *Kepler* target stars are faint and very far from the Sun.

As a point of comparison, the typical Doppler survey target star has a visual magnitude of $V < 9$, while the typical *Kepler* planet-host star is 4–5 magnitudes fainter, corresponding to 40–100 times fainter. The faintness of the *Kepler* planet-host stars makes RV follow-up to measure planet masses either extremely time-consuming or impossible, even with the largest telescopes. Other types of follow-up observations, for e.g., those aimed at studying the atmospheric characteristics of transiting planets, are similarly limited by the faintness of the *Kepler* planet-host stars.

In addition to being faint, *Kepler* targets are also very far from the Sun. The typical distance to a Doppler-detected planet is 30–70 pc, compared to 100–1000 pc for *Kepler* host stars. This means that very few *Kepler* stars have parallax-based distances, which greatly limits the precision with which stellar and planetary radii and masses can be measured, despite *Kepler*'s outstanding photometric precision. While *Kepler* can measure $R_P / R_\star$ to high precision, the precision of R_P is limited by knowledge of $R_\star$, which in turn depends on the star's luminosity and distance (see Section 3.2).

Future space-based transit surveys will benefit from targeting brighter, more proximate stars over a much larger fraction of the sky than currently surveyed by *Kepler*. Such a survey has been selected by NASA as its next Explorer-class mission, called the Transiting Exoplanet Survey Satellite (TESS; Ricker et al., 2010). TESS will be a small satellite in orbit around the Earth, unlike Kepler's Earth-trailing orbit. And rather than staring at a single target field, TESS will use a cluster of four small telescopes to monitor four target fields at a time.

Each TESS telescope has a 7.7-cm aperture, but like ground-based wide-field transit surveys, these small telescopes provide a huge field of view. Compared to *Kepler*'s $100 \deg^2$ target field, each of the four TESS telescopes will stare at a $529 \deg^2$ field of view or a total of $2116 \deg^2$ per pointing, stretching from the celestial pole to the ecliptic. However, TESS is not limited to just one of these target fields. Throughout its mission lifetime TESS will survey the entire sky for planets orbiting bright stars. Based on the results from *Kepler*, we know that there are on average two planets per star with periods detectable by TESS. Thus, the sample of known transiting planets will increase substantially with NASA's next transit survey mission.

6.7 The Future of Microlensing

As described in Chapter 4, past and present microlensing planet surveys have relied on facilities that provide course time sampling (typically one to two observations per star per night), due to their limited fields of view. Future ground-based surveys will be dedicated to searching for the short-duration signals caused by planets and will rely on telescopes with larger fields of view. Additionally, these future surveys will use networks of telescopes distributed around the globe to avoid interruptions from the day–night cycle and weather at any one site.

However, these future, dedicated ground-based surveys will still suffer from several limitations. The first is confusion from crowded fields. Observing toward the galactic bulge provides the large number of targets needed to increase the chances of observing a sizable number of microlensing signals each year. However, ground-based facilities must observe through the Earth's atmosphere, and the crowded target fields will have stars blended together by the effects of seeing. This blending of stars both dilutes the microlensing light curve due to the contaminating light of other stars and increases the difficulty of identifying which stars are responsible for the microlensing signal. Without knowledge of the lens and source star properties, the physical parameters of potential planetary systems cannot be measured reliably.

One solution to these problems is the same as the solution to transit surveys: move the enterprise to outer space. Without the intervening atmosphere to blur the images, the stars can be well separated on the detector and microlensing signals can be traced to specific lens and source stars much more reliably. Space-based observations also benefit from a view of the various target fields that is uninterrupted by the diurnal cycle and weather.

Another solution is to observe at infrared wavelengths, rather than the optical (visual) bandpasses used by current, ground-based surveys. When observing stars at distances of 1–8 kpc, interstellar dust along the line of sight attenuates light from stars. Dust attenuation, referred to by astronomers as *extinction*, is what causes the sun to appear red during sunset and sunrise: shorter wavelengths are scattered by particles in the Earth's atmosphere while redder wavelengths pass through. By observing in infrared wavelengths near 1 μm, future space-based surveys will see through the majority of the Galaxy's dust, thereby providing more targets and more opportunities for microlensing events.

Another potential microlensing technique is to target specific, nearby stars that have known proper motions. Since the star's path along the sky and its angular speed are known, one can examine the nearby star's path and see if it will encounter background stars in the future (Di Stefano et al., 2012). This "pencil-beam" technique has several advantages over traditional wide-field surveys. First, while many known microlensing events involve anonymous, nearly invisible planet-host stars, targeting nearby stars would mean that the host stars would be well characterized. Also, if the lens star is close enough, it's trigonometric parallax can be measured independently of the microlens light curve, and any planet discovered would have an absolute mass and orbit separation, rather than q and s, which are scaled by the Einstein ring angle θ_E (see Section 4.3). Finally, predictions can be made about the exact time that the lensing event will occur, which allows for detailed observational planning and allocation of telescope resources (Lépine & DiStefano, 2012; Sahu et al., 2014).

6.8 The Future of Direct Imaging

Of all the techniques for planet hunting presented in this book, the only method that directly senses light from the planet rather than the star is the imaging technique. While difficult in practice, "taking pictures of planets" is the most powerful method of planet detection, both on an emotional level ("Look, there's a planet, right there") and scientifically. When a planet is directly imaged, it opens up the opportunity to acquire a spectrum of light emitted from the planet. And because the technique is most sensitive to planets in wide orbits, at tens of AU, it offers the opportunity to study planets that reside near their birthplaces. This is in contrast to close-in planets that likely moved inward from where they formed at several to tens of AU, to a fraction of an AU, such as the hot Jupiters.

Indeed, one of the ultimate goals of planet hunting is to discover a planet like our own, in terms of its orbit, physical properties and host star. The foundation for this vision of the future was laid more than 24 years ago when the Voyager spacecraft arrived at the outer reaches of the Solar System. At a distance of about 40 AU from the Sun, Voyager executed a set of commands programmed by Carolyn Porco (University of Arizona) and Candy Hansen (NASA Jet Propulsion Laboratory) that rotated the spacecraft toward the Sun to take a "family portrait" of the Solar System planets. When the frames were later analyzed, there in the nearly overwhelming glare of the Sun was a tiny point of light that later inspired the title of Carl Sagan's book *Pale Blue Dot: A Vision of the Human Future in Space* (1994).

As of now, imaging a similar pale blue dot around another star remains a goal for the distant future, perhaps several decades from now. However, this discovery will not be made as a result of a sudden leap in human capabilities. The groundwork that will lead to this goal is under way right now. The first direct-imaging surveys have discovered massive brown dwarfs and "super Jupiters" orbiting at tens of AU from their host stars. These detections were made using existing, multipurpose instrumentation such as the imagers on the *Hubble* Space Telescope and adaptive optics systems on large, ground-based telescopes such as Keck and the Very Large Telescope (VLT) in Chile. The lessons learned on these first-generation direct imaging surveys have paved the way for the next generation of purpose-built high-contrast imaging systems.

The next generation of direct imaging instruments will take all of the tricks of the trade that have been developed over the past decade and combine them into a single instrument. A good example of this everything-in-one-package approach is the design of the Gemini Planet Imager (GPI; McBride et al., 2011), which is shared by GPI's "little brother" known as *Project 1640* (P1640; Hinkley et al., 2011) on the Palomar Hale 200-inch telescope near San Diego, California. GPI and P1640 are highly ambitious instruments that sit behind powerful adaptive optics systems. P1640's AO system, PALM-3000, has a deformable mirror with 3388 actuators that corrects high-spatial-frequency components of images. Before reaching this high-frequency "tweeter" deformable mirror, the light is first corrected with a separate, 241-actuator low-frequency "woofer" deformable mirror.

The woofer-tweeter AO system feeds an additional calibration unit that can sense residual errors caused not by the Earth's atmosphere, but errors caused by optical aberrations in the AO system. While the AO system senses imperfect sky images, the Cal Unit detects imperfections in the AO-generated image and sends additional corrections to the AO deformable mirrors. The light then passes into an integral-field unit (IFU), in which an array of tiny lenses disperses various spatial regions of the image into spectra. The Spectro-Polarimetric High-contrast Exoplanet REsearch (SPHERE) instrument on the VLT will also have an IFU (Beuzit et al., 2010). Thus, instead of detecting the spatially resolved image of the star and its planet(s) into monochromatic pixels, the pixels contain a spectrum of light. As I described briefly in Chapter 5, the resulting multicolor image allows the color-position-dependent speckles to be separated from the stationary image of a planet. The added benefit is that the planet's spectrum is acquired within the discovery image, allowing for instant characterization of the object.

6.9 Concluding Remarks

In this book I provided an overview of the methods that have led to secure exoplanet detections over the past two decades. The pace of discovery in exoplanetary science is truly remarkable. The first planet orbiting a normal, Sun-like star was detected in 1996. Since then, each year has seen an increase in the numbers of known exoplanets, leading up to our present-day knowledge of thousands of well-characterized planets orbiting a wide variety of stars. Each advancement in technology and improvement in survey strategies has provided an increase in the number of detections as well as the diversity of planetary properties.

As an illustration of the abundance of planetary systems around other stars, consider the statistical analyses of the planet candidates discovered by the *Kepler* Mission. These studies have found that there are one to three planets per star throughout the Galaxy, and that smaller planets like the terrestrial planets of the Solar System outnumber large planets like Jupiter and Saturn by a wide margin (Howard et al., 2010; Fressin et al., 2013). Results like these show quite vividly that the era of planet *hunting* is coming to an end and will soon be replaced by the era of planet *gathering*. The Galaxy is positively teeming with planets, and the challenge before us is to reach out with ever more sensitive instrumentation to harvest the great bounty from the sky.

BIBLIOGRAPHY

Albrow, M., Beaulieu, J.-P., Birch, P., et al. 1998. *Astrophysical Journal*, 509, 687.

Artigau, É., Kouach, D., Donati, J.-F., et al. 2014. *Proceedings of SPIE*, 9147, 914715.

Bakos, G., Noyes, R. W., Kovács, G., et al. 2004. *Publications of the Astronomical Society of the Pacific*, 116, 266.

Basri, G., Borucki, W. J., & Koch, D. 2005. *New Astronomy Review*, 49, 478.

Beaulieu, J.-P., Bennett, D. P., Fouqué, P., et al. 2006. *Nature*, 439, 437.

Beuzit, J.-L., Boccaletti, A., Feldt, M., et al. 2010. In *Pathways towards Habitable Planets*. Astronomical Society of the Pacific. 430, 231.

Black, D. C. 1995. *Annual Review of Astronomy and Astrophysics*, 33, 359.

Bond, I. A., Abe, F., Dodd, R. J., et al. 2001. *Monthly Notices of the Royal Astronomical Society*, 327, 868.

Bouchy, F., Udry, S., Mayor, M., et al. 2005. *Astronomy and Astrophysics*, 444, L15.

Campbell, B., & Walker, G.A.H. 1979. *Publications of the Astronomical Society of the Pacific*, 91, 540.

Carter, J. A., Yee, J. C., Eastman, J., Gaudi, B. S., & Winn, J. N. 2008. *Astrophysical Journal*, 689, 499.

Charbonneau, D., Brown, T. M., Latham, D. W., & Mayor, M. 2000. *Astrophysical Journal Letters*, 529, L45.

Cosentino, R., Lovis, C., Pepe, F., et al. 2012. *Proceedings of SPIE*, 8446, 84461V.

Crepp, J. R., Johnson, J. A., Fischer, D. A., et al. 2012. *Astrophysical Journal*, 751, 97.

Crossfield, I. J. M., Petigura, E., Schlieder, J., et al. 2015. arXiv:1501.03798.

Dawson, R. I., & Johnson, J. A. 2012. *Astrophysical Journal*, 756, 122.

Dawson, R. I., Johnson, J. A., Morton, T. D., et al. 2012. *Astrophysical Journal*, 761, 163.

Di Stefano, R., & Esin, A. A. 1995. *Astrophysical Journal Letters*, 448, L1.

Di Stefano, R., Matthews, J., & Lepine, S. 2012. arXiv:1202.5314.

Doyle, L. R., Carter, J. A., Fabrycky, D. C., et al. 2011. *Science*, 333, 1602.

Dressing, C. D., & Charbonneau, D. 2013. *Astrophysical Journal*, 767, 95.

Dumusque, X., Pepe, F., Lovis, C., et al. 2012. *Nature*, 491, 207.

Dyson, F. W., Eddington, A. S., & Davidson, C. 1920. *Royal Society of London Philosophical Transactions Series A*, 220, 291.

Einstein, A. 1936. *Science*, 84, 506.

Einstein Online. http://www.einstein-online.info/spotlights/grav_lensing_history.

Fischer, D. A., Laughlin, G., Butler, P., et al. 2005. *Astrophysical Journal*, 620, 481.

Fischer, D. A., Marcy, G. W., & Spronck, J.F.P. 2014. *Astrophysical Journal Supplement*, 210, 5.

Fischer, D. A., & Valenti, J. 2005. *Astrophysical Journal*, 622, 1102.

Ford, E. B. 2009. *New Astronomy*, 14, 406.

Fressin, F., Torres, G., Charbonneau, D., et al. 2013. *Astrophysical Journal*, 766, 81.

Gaudi, B. S. 2011. *Exoplanets*, edited by S. Seager. University of Arizona Press, pp. 79–110, 79.

—— 2012. *Annual Review of Astronomy and Astrophysics*, 50, 411.

Gonzalez, G. 1997. *Monthly Notices of the Royal Astronomical Society*, 285, 403.

Gould, A., & Loeb, A. 1992. *Astrophysical Journal*, 396, 104.

Hearnshaw, J. B., Abe, F., Bond, I. A., et al. 2006. 9th Asian-Pacific Regional AU Meeting, 272.

Henry, G. W., Marcy, G. W., Butler, R. P., & Vogt, S. S. 2000. *Astrophysical Journal Letters*, 529, L41.

Hinkley, S., Oppenheimer, B. R., Zimmerman, N., et al. 2011. *Publications of the Astronomical Society of the Pacific*, 123, 74.

Howard, A. W., Marcy, G. W., Johnson, J. A., et al. 2010. *Science*, 330, 653.

Howell, S. B., Sobeck, C., Haas, M., et al. 2014. *Publications of the Astronomical Society of the Pacific*, 126, 398.

Hubble, E. P. 1936. *Realm of the Nebulae*. Yale University Press, 1936.

Jayawardhana, R. 2011. *Strange New Worlds: The Search for Alien Planets and Life beyond Our Solar System*. Princeton University Press.

Johnson, J. A. 2009. *Publications of the Astronomical Society of the Pacific*, 121, 309.

Johnson, J. A., Marcy, G. W., Fischer, D. A., et al. 2006. *Astrophysical Journal*, 647, 600.

Johnson, J. A., Winn, J. N., Cabrera, N. E., & Carter, J. A. 2009. *Astrophysical Journal Letters*, 692, L100.

Johnson, J. A., Winn, J. N., Narita, N., et al. 2008. *Astrophysical Journal*, 686, 649.

Kasting, J. F., 2010. *How to Find a Habitable Planet*. Princeton University Press.

Kasting, J. F., Whitmire, D. P., & Reynolds, R. T. 1993. *Icarus*, 101(1), 108.

Knutson, H. A., Charbonneau, D., Cowan, N. B., et al. 2009. *Astrophysical Journal*, 703, 769.

Kuhn, T. S. 1957. *The Copernican Revolution*. Cambridge: Harvard University Press, 1957.

Latham, D. W., Stefanik, R. P., Mazeh, T., Mayor, M., & Burki, G. 1989. *Nature*, 339, 38.

Lemonick, M. 1998. *Other Worlds: The Search for Life in the Universe*. Simon & Schuster.

Lemonick, M. 2012. *Mirror Earth: The Search for Our Planet's Twin*. Walker & Company.

Lépine, S., & DiStefano, R. 2012. *Astrophysical Journal Letters*, 749, L6.

Lissauer, J. J., Fabrycky, D. C., Ford, E. B., et al. 2011. *Nature*, 470, 53.

Mahadevan, S., Ramsey, L., Bender, C., et al. 2012. *Proceedings of SPIE*, 8446, 84461S.

Mandel, K., & Agol, E. 2002. *Astrophysical Journal Letters*, 580, L171.

Mao, S., & Paczynski, B. 1991. *Astrophysical Journal Letters*, 374, L37.

Marcy, G. W., & Butler, R. P. 1996. *Astrophysical Journal Letters*, 464, L147.

Mayor, M., Bonfils, X., Forveille, T., et al. 2009. *Astronomy and Astrophysics*, 507, 487.

Mayor, M., Pepe, F., Queloz, D., et al. 2003. *The Messenger*, 114, 20.

Mayor, M., & Queloz, D. 1995. *Nature*, 378, 355.

Mazeh, T., Naef, D., Torres, G., et al. 2000. *Astrophysical Journal Letters*, 532, L55.

McBride, J., Graham, J. R., Macintosh, B., et al. 2011. *Publications of the Astronomical Society of the Pacific*, 123, 692.

Morton, T. D., & Johnson, J. A. 2011. *Astrophysical Journal*, 738, 170.

Morton, T. D., & Swift, J. 2014. *Astrophysical Journal*, 791, 10.

Naef, D., Latham, D. W., Mayor, M., et al. 2001. *Astronomy & Astrophysics*, 375, L27.

Oppenheimer, B. R., & Hinkley, S. 2009. *Annual Review of Astronomy & Astrophysics*, 47, 253.

Origlia, L. 2013. SF2A-2013: *Proceedings of the Annual Meeting of the French Society of Astronomy and Astrophysics*, p. 287.

Orosz, J. A., Welsh, W. F., Carter, J. A., et al. 2012. *Science*, 337, 1511.

Paczynski, B. 1991. *Astrophysical Journal Letters*, 371, L63.

Penny, A. 2011. *Astronomy and Geophysics*, 52(1), 21.

Pepe, F., Molaro, P., Cristiani, S., et al. 2014. *Astronomische Nachrichten*, 335, 8.

Perryman, M. 2011. *The Exoplanet Handbook*. Cambridge University Press.

Pogge, R. 2005. "A Brief Note on Religious Objections to Copernicus." http://www.astronomy.ohio-state.edu/~pogge/Ast161/Unit3/response.html.

Pont, F., Sing, D. K., Gibson, N. P., et al. 2013. *Monthly Notices of the Royal Astronomical Society*, 432, 2917.

Pollacco, D. L., Skillen, I., Collier Cameron, A., et al. 2006. *Publications of the Astronomical Society of the Pacific*, 118, 1407.

Quintana, E. V., Barclay, T., Raymond, S. N., et al. 2014. *Science*, 344, 277.

Quirrenbach, A., Amado, P. J., Caballero, J. A., et al. 2014. *Proceedings of SPIE*, 9147, 91471F.

Ricker, G. R., Winn, J. N., Vanderspek, R. K., et al. 2015. *Journal of Astronomical Telescopes*, Instruments, and Systems, 1(1).

Sagan, C. 1994. *Pale Blue Dot: A Vision of the Human Future in Space*. Random House.

Sahu, K. C., Bond, H. E., Anderson, J., & Dominik, M. 2014. arXiv:1401.0239.

Santos, N. C., Israelian, G., & Mayor, M. 2004. *Astronomy and Astrophysics*, 415, 1153.

Sato, B., Fischer, D. A., Henry, G. W., et al. 2005. *Astrophysical Journal*, 633, 465.

Schilling, G. 1996. *Science*, 273, 429.

Seager, S. 2010. *Exoplanet Atmospheres: Physical Processes*. Princeton University Press.

Seager, S., ed. 2011. *Exoplanets*. University of Arizona Press, p. 526.

Seager, S., Deming, D., & Valenti, J. A. 2009. In *Astrophysics in the Next Decade*. Springer, p. 123.

Seager, S., & Mallén-Ornelas, G. 2003. *Astrophysical Journal*, 585, 1038.

Shields, A. L., Meadows, V. S., Bitz, C. M., et al. 2013. *Astrobiology*, 13, 715.

Sobel, D. 2011. *A More Perfect Heaven*. Walker Books, 2011.

Swift, J. J., Bottom, M., Johnson, J. A., et al. 2014. arXiv:1411.3724.

Swift, J. J., Johnson, J. A., Morton, T. D., et al. 2013. *Astrophysical Journal*, 764, 105.

Struve, O. 1952. *The Observatory*, 72, 199.

Tarter, J. 2001. *Annual Review of Astronomy and Astrophysics*, 39, 511.

Tull, R. G. 1998. *Proceedings of SPIE*, 3355, 387.

Udalski, A. 2003. *Acta Astronomica*, 53, 291.

Vanderburg, A., & Johnson, J. A. 2014. *Publications of the Astronomical Society of the Pacific*, 126, 948.

Vanderburg, A., Montet, B. T., Johnson, J. A., et al. 2014. arXiv:1412.5674.

Vogt, S. S., Allen, S. L., Bigelow, B. C., et al. 1994. *Proceedings of SPIE*, 2198, 362.

Vogt, S. S., Radovan, M., Kibrick, R., et al. 2014. *Publications of the Astronomical Society of the Pacific*, 126, 359.

Walker, G.A.H. 2012. *New Astronomy Review*, 56, 9.

Wisdom, J., & Holman, M. 1991. *Astronomical Journal*, 102, 1528.

Wolszczan, A. 1995. In *Millisecond Pulsars: A Decade of Surprise*. Astronomical Society of the Pacific. 72, 377.

Wolszczan, A., & Frail, D. A. 1992. *Nature*, 355, 145.

Wright, J. T. & Howard, A. W. 2009. *Astrophysical Journal Supplement*, 182, 205.

Wright, J. T., Marcy, G. W., Howard, A. W., et al. 2012. *Astrophysical Journal*, 753, 160.

GLOSSARY

Adaptive Optics (AO): A technology used to improve the performance of optical systems by reducing the effect of wavefront distortions by correcting the deformations of an incoming wavefront by deforming a mirror in order to compensate for the distortion.

Anomaly: The three angular parameters of a Keplerian orbit are known as anomalies, including the eccentric anomaly, mean anomaly, and true anomaly. All three need to be specified in order to predict the position of an object in a two-body orbit.

Apastron: The point of greatest separation of a star and an orbiting body with an eccentric orbit.

Aperture: The size of the opening through which light passes in an optical instrument such as a telescope. For circular optics, the size of the aperture is typically given as the diameter.

Aphelion: The point in the orbit of a planet or other celestial body where it is farthest from the Sun.

Argument of Periastron: A parameter of orbital motion describing the orientation of the orbit, given by the symbol ω. Specifically, the argument of periastron is the angle from the ascending node (one of the points of intersection between the orbit ellipse and the plane of reference) to periastron, measured in the direction of motion.

Astrometry: The branch of astronomy that involves precise measurements of the positions and movements of stars and other celestial bodies.

Astronomical Unit (AU):   An astronomical unit of measure equal to the average distance between the Earth and the Sun, approximately 1.5×10^{13} cm.

Blueshift: A Doppler shift in the lines of an object's spectrum toward bluer wavelenths, indicating that an object is moving toward the observer. The larger the blueshift, the faster the object is moving.

Doppler Effect: An increase (or decrease) in the frequency of sound, light, or other wave phenomenon as the source and observer move toward (or away from) each other.

Eccentricity: The measure of how an object's orbit differs from a circle, with higher eccentricities corresponding to orbits that are more elongated.

Eclipse: The total or partial blocking of one celestial body by another. Typically used for the passage of one star in a stellar binary in front of the other star.

Egress: The passage of an eclipsing (transiting) body across the limb of the central (orbited) body occurring just after full eclipse (transit).

Epicycle: A small circle whose center moves around the circumference of a larger one. Historically, an epicycle was used to describe planetary orbits in the Ptolemaic system.

Exoplanet: A planet orbiting a star other than the Sun, with otherwise the same defining characteristics of a Solar System planet.

General Relativity: Also known as the general theory of relativity, it is the geometric theory of gravitation published by Albert

Einstein in 1915 and the current description of gravitation in modern physics.

Hydrostatic Equilibrium: A state that occurs in stars and other astronomical bodies when the inward pull of gravity is balanced by internal pressure.

Impact Parameter: The perpendicular distance, or closest approach, between the path of a moving object and the center of a separate object that the moving object is approaching.

Infrared: Corresponds to light having wavelengths longer than visible light and shorter than microwave, spanning roughly 0.8 microns ("near" infrared) to 1 mm ("far" infrared).

Ingress: The passage of an eclipsing (transiting) body across the limb of the central (orbited) body occurring just before full eclipse (transit).

Kelvin: A temperature scale used in astronomy given by the symbol K. The freezing point of water corresponds to 273 K, and 0 K is known as absolute zero temperature.

Kepler's Laws: *The Law of Orbits*—all planets move in elliptical orbits, with the sun at one focus. *The Law of Areas*—a line that connects a planet to the sun sweeps out equal areas in equal times. *The Law of Periods*—the square of the period of any planet is proportional to the cube of the semi-major axis of its orbit.

Kiloparsec: An astronomical distance equal to 1000 parsecs.

Limb Darkening: An optical effect seen in stars, where the center part of the disk appears brighter than the edge or limb of the image.

Line of Sight: A straight line along which an observer views an object.

Luminosity: A measure of the power output of a star, with units of energy per time.

Main Sequence: The locus in the stellar temperature-luminosity diagram (Hertzsprung-Russell diagram) along which hydrogen-burning stars lie. The Sun is currently on the main sequence.

Magnitude: A way of expressing the brightness of a star. Brighter stars have smaller magnitude values than fainter stars. The brightest star in the night sky has a magnitude of roughly -1.4, and the faintest magnitude visible to humans is 6. Each integer interval corresponds to roughly a factor of 2.51 in brightness.

Metallicity: The measure of the amount of heavy elements, or "metals" in a star, where metal corresponds to elements heavier than helium in the vernacular of astronomers.

Parallax: The apparent change in position of two objects viewed from different locations.

Parsec: A unit of distance often used in astronomy corresponding to the distance at which an object moves 1 arcsecond when the observer moves by 1 AU.

Periastron: The point of closest separation of a star and an orbiting body with an eccentric orbit.

Perihelion: The point in the orbit of a planet or other celestial body where it is closest to the Sun.

Photometry: The study of astronomical objects by measuring the flux of the object's emitted radiation.

Photon: A quantum particle of light comprising a certain amount of electromagnetic energy.

Photosphere: The bright apparent surface of the Sun. The photosphere is not a physical surface. Rather it corresponds to the point at which photons can escape without scattering or absorption.

Plage: Bright regions on a star's surface.

Planet: The International Astronomical Union (IAU) defines a planet as a Solar System body that meets three criteria: (1) is in orbit around the Sun, (2) has sufficient mass for its self-gravity to overcome rigid body forces so that it assumes a hydrostatic equilibrium (nearly round) shape, and (3) has cleared smaller objects from its gravitational neighborhood.

Protoplanetary Disk: A flattened, rotating distribution of gas and dust surrounding a young newly formed star. It is thought that planets are eventually formed from the gas and dust within the protoplanetary disk.

Pulsar: A rotating neutron star that emits energy along its magnetic field axis. As the rotation of the star causes this axis to sweep along the line of sight, the observer sees regular pulses of light.

Radial Velocity: The velocity of an object projected along the observer's line of sight.

Redshift: A Doppler shift in the lines of an object's spectrum toward redder wavelenths, indicating that an object is moving away from the observer. The larger the redshift, the faster the object is moving.

Retrograde Motion: The phenomenon by which a celestial body appears to slow down, stop, then move in the opposite direction with respect to the stars in the night sky.

Semimajor Axis: The longest radius of an eccentric orbit, lying along a line connecting apastron and periastron.

Semiminor Axis: The shortest radius of an ellipse, lying along a line perpendicular to the semimajor axis running through the geometric center of the ellipse.

Spectrometer: An instrument that separates light from an astronomical source into different wavelengths, producing a spectrum.

Spectroscopy: The technique of observing the spectra of astronomical objects in order to measure or infer their physical characteristics, such as chemical composition and radial velocity.

Spectrum: A band of colors, as seen in a rainbow, produced by separation of the components of light by their different degrees of refraction according to wavelength.

Star: A spherical, luminous collection of gas in a state of hydrostatic equilibrium, with luminosity generated by internal nuclear fusion or thermal radiation.

Starspot: Regions on a star's surface that are cooler and darker than the surrounding stellar surface.

Transient: An astronomical phenomenon that occurs unpredictably with a duration ranging from seconds to days, weeks, or even years. Examples include supernova explosions and gamma ray bursts.

Transit: The passage of a smaller body across the disk of a larger one, as in the eclipse of a star by a planet.

INDEX